ICPC-3
INTERNATIONAL
CLASSIFICATION OF
PRIMARY CARE

WONCA Family Medicine

About the Series

The WONCA Family Medicine series is a collection of books written by world-wide experts and practitioners of family medicine, in collaboration with The World Organization of Family Doctors (WONCA).

WONCA is a not-for-profit organization and was founded in 1972 by member organizations in 18 countries. It now has 118 Member Organizations in 131 countries and territories with membership of about 500,000 family doctors and more than 90 per cent of the world's population.

Every Doctor: Healthier Doctors = Healthier Patients
Leanne Rowe, Michael Kidd

Family Practice in the Eastern Mediterranean Region: Primary Health Care for Universal Health Coverage
Hassan Salah, Michael Kidd

Family Practice in the Eastern Mediterranean Region: Universal Health Coverage and Quality Primary Care
Hassan Salah, Michael Kidd

Family Medicine: The Classic Papers
Michael Kidd, Iona Heath, Amanda Howe

Global Primary Mental Health Care: Practical Guidance for Family Doctors
Christopher Dowrick

How To Do Primary Care Educational Research: A Practical Guide
Mehmet Akman, Valerie Wass, Felicity Goodyear-Smith

How To Do Primary Care Research
Felicity Goodyear-Smith, Bob Mash

ICPC- 3 International Classification of Primary Care: User Manual and Classification
Kees van Boven, Huib Ten Napel

International Perspectives on Primary Care Research
Felicity Goodyear-Smith, Bob Mash

Migrant Health: A Primary Care Perspective
Bernadette N. Kumar, Esperanza Diaz

Primary Health Care Around the World: Recommendations for International Policy and Development
Chris van Weel, Amanda Howe

The Contribution of Family Medicine to Improving Health Systems: A Guidebook from the World Organization of Family Doctors
Michael Kidd

For more information about this series please visit: https://www.crcpress.com/WONCA-Family-Medicine/book-series/WONCA

ICPC-3
INTERNATIONAL CLASSIFICATION OF PRIMARY CARE

User Manual and Classification

EDITED BY

Kees van Boven and Huib Ten Napel

Prepared by the ICPC-3 Consortium of WONCA, the
World Organization of Family Doctors

CRC Press
Taylor & Francis Group
Boca Raton London New York

CRC Press is an imprint of the
Taylor & Francis Group, an **informa** business

Third edition published 2022
by CRC Press
6000 Broken Sound Parkway NW, Suite 300, Boca Raton, FL 33487-2742

and by CRC Press
2 Park Square, Milton Park, Abingdon, Oxon, OX14 4RN

© 2022 WONCA

Previously published by Oxford University Press

CRC Press is an imprint of Taylor & Francis Group, LLC

ISBN: 978-1-03-205343-1 (hbk)
ISBN: 978-1-03-205339-4 (pbk)
ISBN: 978-1-00-319715-7 (ebk)

DOI: 10.1201/9781003197157

Typeset in Minion Pro
by Newgen Publishing UK

Contents

Please note this book is accompanied by a free User Manual to guide readers in the implementation of ICPC-3. This can be downloaded from https://www.routledge.com/ICPC-3-International-Classification-of-Primary-Care-User-Manual-and-Classification/Boven-Napel/p/book/9781032053394

Foreword by Donald Li

The publication of ICPC-3 is a very welcome development for primary care and family medicine globally. As we strive towards the achievement of the Sustainable Development Goals generally and universal health coverage specifically, we who are working every day in primary care understand that reflecting on our work through a structured coding or classification system is not feasible when based on diagnoses alone. Patient interaction with primary care professionals is not always based on a definitive diagnosis, but is often based on a symptom or a series of symptoms which curtail normal life for the patient.

In primary care, we understand that the most important issue is the reason for the encounter between the patient and the primary care team. The reason for an encounter could relate to illness prevention, to health promotion, to immunisation and vaccination programmes, to a range of signs and symptoms which are worrying the patient, or to ongoing management of a chronic health problem. The skill in primary care is managing the whole patient, rather than their separate diagnoses. Managing uncertainty is a key element of delivering primary care, where a diagnosis can be hard to assign.

ICPC-3 allows patients' health issues to be tracked over time. It shows both the frequency and the distribution of health issues commonly encountered in primary care, and, importantly, it reflects the way the primary care team addresses issues and solves problems. The classification system is easy for practitioners to use. Importantly, the aggregation, collation and analysis of data is useful for primary care research purposes, allowing exchange of information with policymakers, managers, and funding agencies at local, national and global levels.

ICPC-3 allows us to reflect, in a realistic way, what is happening in primary care to address the delivery of comprehensive, coordinated, continuous, community-based care. Used widely, ICPC-3 will not only show what is being done to achieve universal health coverage, but also help us to identify where there are gaps in our primary care systems and contribute to improvements in the delivery of care.

WONCA welcomes the adoption of ICPC-3 as the preferred tool to classify and code primary care activity across the globe. We are proud of our colleagues who have worked on this development so assiduously, skillfully led by Kees van Boven and Huib Ten Napel. We look forward to a near future when what we do on an everyday basis with and for our patients is realistically reflected using a coding and classification system customised to real primary care provision.

Donald Li
President of WONCA

Acknowledgements

CONTRIBUTIONS

The development of the ICPC-3 has been made possible by financial and expert support from the following ICPC-3 Consortium Members.

ICPC-3 Consortium Steering Group Members

Chris van Weel, Chair	WONCA / Radboud University Medical Centre, Nijmegen, Department of Primary and Community Care, Netherlands
Garth Manning	WONCA CEO
Harris Lygidakis	WONCA CEO
An De Sutter	Ghent University / Belgium
Gustavo Gusso	Socidade Brasileira de Medicina de Família e Comunidade (SBMFC) / Brasil
Heikki Virkkunen	National Institute for Health and Welfare (THL) / Finland
Thierry Dart	Agency for digital health (ANS) / France
Thomas Frese	WONCA Europe
Thomas Kühlein	WICC / Germany
Tjeerd van Althuis	Dutch College / Netherlands

ICPC-3 Consortium Taskforce Members

Olawunmi Olagundoye	WICC / Nigeria
Khing Njoo	Dutch College / Netherlands
Danniel Knupp	Socidade Brasileira de Medicina de Família e Comunidade (SBMFC) / Brasil
Laurent Letrilliart	Agency for digital health (ANS) / France
Härkönen Mikko	National Institute for Health and Welfare (THL) / Finland
Diego Schrans	Ghent University / Belgium

Observers to the ICPC-3 Consortium Taskforce

Bjørn Gjelsvik	Norwegian College of General Practice (NCGP) / Norway
Øystein Hetlevik	Norwegian College of General Practice (NCGP) / Norway
Preben Larsen	Danish College of General Practitioners / Denmark

Robert Jakob	WHO Geneva
Thomas Frese	WONCA Europe

The WONCA International Classification Committee (Since 2011)

Ana Kareli	Georgia
Anders Grimsmo	Norway
Benjamin Fauquert	Belgium
Bjørn Gjelsvik	Norway
Daniel Pinto	Portugal
Diego Schrans, Chair	Belgium
Dimitris Kounalakis*	Greece
Elena Cardillo*	Italy
Ferdinando Petrazzuoli	Italy
Francois Mennerat	France
Gojo Zorz*	Slovenia
Graeme Miller	Australia
Gustavo Gusso	Brazil
Harald Kornfeil	Austria
Heikki Virkkunen	Finland
Heinz Bhend	Switzerland
Helena Britt	Australia
Ines Zelic	Croatia
Jean Karl Soler*	Malta
Julie Gordon	Australia
Kees van Boven	Netherlands
Krishna Mohan	India
Laurent Letrilliart	France
Luciana Tanno	Italy
Marc Jamoulle	France
Marc Verbeke*	Belgium
Marija Vrca Botica	Croatia
Marten Kvist	Finland
Michel de Jonghe	Belgium
Miguel Pizzanelli	Uruguay
Mike Klinkman, Former Chair	USA
Mikko Härkönen	Finland
Nick Booth	United Kingdom
Nicola Buono*	Italy
Olawunmi Olagundoye*	Nigeria
Olesya Vynnyk	Ukraine
Osman Abdulhamid	Sudan
Øystein Hetlevik	Norway
Pauline Boeckxstaens	Belgium
Preben Larsen	Denmark
Ray Simkus	Canada

Sebastian Juncosa	Spain
Shabir Moosa	South Africa
Shinsuke Fujita	Japan
Simone Postma*	Netherlands
Taran Borge	Norway
Thomas Frese	Germany
Thomas Kühlein, Former Chair	Germany
Tuija Savolainen	Finland
Vadum Vus	Ukraine

* WICC taskforce A member

Contributors

Members of the ICPC-3 Consortium Core Group/Secretariat

Kees van Boven	Project manager
Egbert van der Haring	Software
Huib Ten Napel	Project manager
Marc Verbeke	Classification expert

Acronyms

ATC	Anatomic Therapeutic Chemical [classification system]
ATCIF	Arrêts de Travail en médecine générale à partir de la Classification Internationale du Fonctionnement, du handicap et de la santé
CSV	Consent Scale Value
EoC	episode of care
FBV	Facilitator or Barrier Value
FEV	Forced Expiratory Volume
ICD	International Classification of Diseases and Related Health Problems
ICF	International Classification of Functioning, Disability and Health
ICHI	International Classification of Health Interventions
ICHPPC	International Classification of Health Problems in Primary Care
ICPC	International Classification of Primary Care
ID	identification number
NOS	not otherwise specified
OECD	Organisation for Economic Co-operation and Development
PCFS	Primary Care Functioning Scale
RFE	reason for encounter
RFEC	Reason for Encounter Classification
SNOMED CT	Systematic Nomenclature for Medicine – Clinical Terms
UHC	universal health coverage
WHO	World Health Organization
WHO-DAS 2.0	World Health Organization Disability Assessment Scale 2.0
WHO-FIC	World Health Organization Family of International Classifications
WICC	WONCA International Classification Committee
WONCA	World Organization of Family Doctors

CHAPTER 1

Introduction

Welcome to the third version of the International Classification of Primary Care (ICPC).

This manual is intended to give insight into the underlying principles of how and why the ICPC-3 has been built and offers detailed guidance in the use of its contents.

The ICPC-3 is developed in the first place for online electronic application and use. For this purpose, the ICPC-3 is available in an online browser on the ICPC-3 website. The website contains all relevant information on the ICPC-3, including educational material: www.ICPC-3.info.

This manual contains a condensed part of the ICPC-3, without the electronic features offered by the ICPC-3 browser (as explained in Chapter 2).

OVERVIEW OF THE ICPC-3

The content of the classification has changed, and it now has a Framework and contains new chapters.

- The classification has a Framework that underlines the importance of interrelations between all chapters of the ICPC from a person-centred perspective.
- The classification has a systematic list with a new structure for the sequence of chapters:
  - a new chapter entitled Visits for general examination, routine examination, family planning, prevention and other visits, for non-problem-related reasons for encounter and episodes
  - chapters on body/organ systems have new components relating to Symptoms, complaints and abnormal findings, and Diagnoses and diseases
  - a new chapter on Social problems, covering social and environmental factors
  - a chapter headed Interventions and processes, subdivided into Diagnostic and monitoring interventions, Therapeutic and preventive interventions, Programmes related to reported conditions (a new component), Results, Consultation, referral and other reasons for encounter, and Administrative
  - a new chapter entitled Functioning, consisting of Activities and participation, and Functions
  - a new chapter called Functioning related, covering Environmental factors and Personality functions
  - a new chapter entitled Regional extensions, with national or regional classes

DOI:10.1201/9781003197157-1

- a new chapter entitled Emergency codes, on codes for emergency use with epidemiological importance in relation to risk of (national or international) spreading of infections
- a chapter called Extension codes, covering codes provided as supplementary codes or additional positions to give more detail or meaning to the initial code, if so desired
- The codes have been expanded from three to four digits, giving more scope for additional classes and for corrections of classification of classes. Along with the two new components in the chapters on body/organ systems, this new structure allows for new demands to be addressed in future updates.

ACCEPTANCE OF THE ICPC-3

The Executive response is as follows:

- After consideration of the proposals prepared by the ICPC-3 Consortium members and brought forward by the ICPC-3 Steering Group on the International Classification for Primary Care – Third Edition, WONCA executives ACCEPTED and ENDORSED the ICPC-3 on 16 April, 2021.
- The Executive RECOMMENDS the use and implementation of the full ICPC-3 for all primary health care professionals on a global scale.
- The Executive REQUESTS the ICPC-3 Consortium publishes the ICPC-3 manual.

HISTORY OF THE ICPC[i]

Until the mid-1970s, most morbidity data collected in primary care research were classified using the International Classification of Diseases (ICD).[1,2]

This had the important advantage of international recognition, aiding comparability of data from different countries. However, there was the disadvantage that the many symptoms and non-disease conditions that were present in primary care were difficult to code with the ICD, originally designed for application to mortality statistics and with a disease-based structure.

Recognising the problems of the ICD and the need for an internationally recognised classification for general practice, the WONCA Classification Committee designed the International Classification of Health Problems in Primary Care (ICHPPC), first published in 1975[3] and with a second edition in 1979[4] related to the ninth revision of the ICD. Although this provided a section for the classification of some undiagnosed symptoms, it was still based on the ICD structure and remained inadequate. A third edition in 1983 added to its criteria for the use of most of the classes,[5] greatly adding to the reliability with which it could be used but not overcoming its deficiencies for primary care. A new classification was needed for both the patient's reason for encounter (RFE) and the provider's record of the patient's problems.

At the 1978 World Health Organization (WHO) International Conference on Primary Health Care in Alma Ata,[6] adequate primary health care was recognised as the key to the goal of 'health for all by the year 2000'. Subsequently, both WHO and WONCA recognised that the building of appropriate primary care systems to allow the assessment and implementation of health care priorities was only possible if the right information was available to health care planners. This led to the development of new classification systems.

Later in 1978, WHO appointed what became the WHO Working Party for Development of an International Classification of Reasons for Encounter in Primary Care.[7] This group, most of whose members were also members of the WONCA Classification Committee, developed the Reason for Encounter Classification (RFEC),[7,8,9] which later became the ICPC.

An RFE is the agreed statement of the reason(s) why a patient enters the health care system, representing the demand for care by that person. This may be symptoms or complaints (e.g. headache or fear of cancer), a known disease (e.g. flu or diabetes), a request for preventive or diagnostic services (e.g. a blood pressure check or an ECG), a request for treatment (e.g. a repeat prescription), to get test results, or an administrative purpose (e.g. to get a medical certificate). These reasons are usually related to one or more underlying problems that the doctor formulates at the end of the encounter as the conditions that have been treated, which may or may not be the same as the RFE.

Disease classifications are designed to allow the health care provider's interpretation of a patient's health care problem to be coded in the form of an illness, disease or injury. In contrast, the RFEC focuses on data elements from the patient's perspective.[7,10,11,12] In this respect, it is patient oriented rather than disease oriented or provider oriented. The RFE, or demand for care, given by the patient has to be clarified by the physician or other primary care health worker before there is an attempt to interpret and assess the patient's health problem in terms of a diagnosis or to make any decision about the process of management and care.

The working group developing the RFEC tested several versions in field trials. In the course of this feasibility testing, it was noted that the RFEC could easily be used to classify simultaneously the RFEs and two other elements of problem-oriented care: the process of care and the health problems diagnosed. Thus, this conceptual framework allowed for the evolution of the RFEC into the ICPC.

Problems in relation to the concurrent development of the ICD-10 prevented WHO from publishing the RFEC. However, WONCA was able to use it to develop the ICPC and published the first edition in 1987.[13] While the ICPC-1 was much more appropriate for primary care than previous classifications based on the ICD framework, it did not provide inclusion criteria for the classes or any cross-referencing. It was, in this respect, less useful than the previous publication, ICHPPC-2-Defined, though it referred to the latter as a source of inclusion criteria.

In 1980 WONCA became a non-government organisation in official relations with WHO, and joint work since then has led to a better understanding of the requirements of primary care for its own information systems and classifications within an overall framework encompassing all health services.

In 1985 a project began in several European countries to use the new classification system to produce morbidity data from general practice for national health information systems. This involved translations of the classification and comparative studies across countries. The results were published in 1993 in a book including an update of the ICPC.[14]

ICPC-1

The first edition of the ICPC broke new ground in the world of classification when it was published in 1987 by WONCA, the World Organization of National Colleges, Academies, and Academic Associations of General Practitioners/Family Physicians, now known as the World Organization of Family Doctors. For the first time, health care providers could classify, using a single classification, three important elements of the health care encounter: RFEs; diagnoses or problems; and process of care. Linkage of elements permitted categorisation from the beginning of an encounter to its conclusion.

The new classification departed from the traditional ICD chapter format in which the axes of several chapters vary from body systems (Chapters III, IV, V, VI, VII, VIII, IX, X, XI, XIII and XIV) to aetiology (Chapters I, II, XVII, XIX, XX) to others (Chapters XV, XVI, XVIII, XXI). This mixture of axes created confusion, since diagnostic entities could, with equal logic, be classified in more than one chapter; for example, influenza could be classified in the infections chapter, the respiratory chapter or both. Instead of conforming to this format, most of the ICPC chapters were based on body systems, following the principle that localisation has precedence over aetiology. Components that were part of each chapter – the RFEs, interventions/processes of care, diagnoses or problems – permitted considerable specificity for all three elements of the encounter, and their symmetrical structure and frequently uniform numbering across all chapters facilitated usage even in manual recording systems. The rational and comprehensive structure of the ICPC was a compelling reason to consider the classification a model for future international classifications. The ICPC was regarded as a biaxial classification or, in medical informatics terms, a second-generation classification.

Since publication, the ICPC has gradually received increasing recognition worldwide as an appropriate classification for general/family practice and primary care, and it has been used extensively in some parts of the world, notably Europe[14] and Australia.[15]

ICPC-2[16] and ICPC-2-R[17]

The second edition of the ICPC was prepared for two main reasons: to relate it to the tenth edition of the ICD (ICD-10), published by WHO in 1992,[2] and to add inclusion criteria and cross-referencing for many of the classes.

In the interests of stability and consistency, very few changes were made to the classification, though many were suggested and were the subject of ongoing work by the WONCA Classification Committee. At the same time, the second edition included information about new developments in the conceptual basis of understanding general/family practice which have arisen in large part from the use of a classification appropriate to the discipline.

Immediately after the publication of the ICPC-2-R, it was agreed that there was a need for a major revision of the ICPC-2. Changes in the way family medicine, and medicine in general, were conceptualised and recorded required more radical revision than yearly updates could reasonably handle, and the very structure of the ICPC-2 needed to be adjusted or changed. Expanding the ICPC with classes on functioning, next to the need for new classes on non-episode-related information, was an ongoing subject of discussion.

ICPC-3

In the past 10 to 15 years, the desire to develop the ICPC-3 dominated the agenda for WONCA International Classification Committee (WICC) meetings. A major step forward in 2017 was the creation of a consortium of colleges of family medicine and interested national organisations, in collaboration with WONCA and led by the University of Nijmegen, to fund the work involved. In January 2018 the development project started in the first instance by setting the Framework for the new ICPC-3, based on the most recent principles of classification development and taking into account a variety of user needs. The project was completed within a time frame of 3 years, resulting in the launch of the ICPC-3 by WONCA president Donald Li on 15 December, 2020.

ICPC and ICD

The ICPC has always been linked with the well-known and widely used ICD, published by WHO.[18] The first edition of the ICPC contained a list of conversion codes to link to the ICD-9. Since then, the ICD-10 was introduced, and the ICPC-2 was carefully mapped to the ICD-10 so that conversion systems could be used. Extensive empirical research has confirmed that the ICPC and the ICD are complementary rather than in competition. The ICPC deals with the complexity of primary health care in all relevant settings in a comprehensive manner. The ICD mainly serves for international comparability of mortality data and for national morbidity data, mainly in hospital settings. The ICPC-3 is the cement or glue between the different health care settings and versions of classifications used. In the ICPC-3, important concepts for primary health care have therefore been linked to the ICD-10 and the ICD-11. From about 9,000 meaningful terms (in the Thesaurus), 2,900 are linked to the ICD-10 and the ICD-11 and 4,560 to terms in the Systematic Nomenclature for Medicine – Clinical Terms (SNOMED CT). The linkages are based on what are important, frequently used and meaningful concepts.

These relations are further explained in Chapter 4.

NOTE
i   Part of the history of the ICPC given here is identical to the history in the ICPC-2-R.

REFERENCES
1.   International Classification of Diseases, Ninth Revision (ICD-9). Geneva, World Health Organization, 1977.
2.   International Statistical Classification of Diseases and Related Health Problems, Tenth revision (ICD-10). Geneva, World Health Organization, 1992.
3.   International Classification of Health Problems in Primary Care (ICHPPC). Chicago, World Organization National Colleges, Academies and Academic Associations of General Practitioners/Family Physicians (WONCA)/American Hospital Association (AHA), 1975.

4. ICHPPC-2 (International Classification of Health Problems in Primary Care). Oxford, Oxford University Press, 1979.

5. ICHPPC-2-Defined: International Classification of Health Problems in Primary Care. 3rd ed. Oxford, Oxford University Press, 1983.

6. Report of the International Conference on Primary Health Care, Alma Ata, USSR, 6–12 September 1978. Geneva, World Health Organization.

7. Meads S. The WHO Reason for Encounter Classification. *WHO Chronicle* 1983; 37 (5): 159–162.

8. Lamberts H, Meads S, Wood M. Classification of reasons why persons seek primary care: pilot study of a new system. *Public Health Rep* 1984; 99: 597–605.

9. Lamberts H, Meads S, Wood M. Results of the international field trial with the Reason for Encounter Classification (RFEC). *Med Sociale Preventive* 1985; 30: 80–87.

10. Working Party to Develop a Classification of the 'Reasons for Contact with Primary Health Care Services'. Report to the World Health Organization, Geneva, Switzerland, 1981.

11. Wood M. Family medicine classification systems in evolution. *J Fam Pract* 1981; 12: 199–200.

12. Olde Hartman T, van Ravesteijn H, van Boven K, van Weel-Baumgarten E, van Weel C. Why the 'reason for encounter' should be incorporated in the analysis of outcome of care. *Br J Gen Pract* 2011; 61 (593): e839–e841.

13. Lamberts H, Wood M (eds.). *ICPC: International Classification of Primary Care.* Oxford, Oxford University Press, 1987.

14. Lamberts H, Wood M, Hofmans-Okkes I (eds.). *The International Classification of Primary Care in the European Community: with Multi-Language Layer.* Oxford, Oxford University Press, 1993.

15. Bridges-Webb C, Britt H, Miles DA, Neary S, Charles J, Traynor V. Morbidity and treatment in general practice in Australia 1990–1991. *Med J Aust* 1992; 157, Suppl. 19 Oct: S1–S56.

16. ICPC-2, International Classification of Primary Care. Prepared by the Classification Committee of WONCA. Oxford, Oxford University Press, 1998.

17. International Classification of Primary Care. ICPC-2-R. Oxford, Oxford University Press, 2005.

18. Wood M, Lamberts H, Meijer JS, Hofmans-Okkes IM. The conversion between ICPC and ICD-10: requirements for a family of classification systems in the next decade. In: Lamberts H, Wood M, Hofman-Okkes I (eds.), *The International Classification of Primary Care in the European Community: with Multi-Language Layer.* Oxford, Oxford University Press, 1993: 18–24.

CHAPTER 2

Basic Principles
How the ICPC-3 Was Built

Nowadays, classifications are used for multiple purposes and in different ways. There is a need for electronic formats for use in electronic health records; electronic versions for desktop, tablet and mobile phone use; and paper-based formats, such as books and condensed overviews. The classes must be easy to find, which requires that the content is represented in the format of an interface terminology or thesaurus. In addition, the data encoded with the classification must be interchangeable with other classifications and terminologies.

To achieve this content, the ICPC-3 is:

- based on a unifying Framework to describe the context
- builds on a Content Model, which describes the properties for the content and the maintenance attributes
- subject to a dedicated review process

Before starting the review process, the ICPC-3 Framework and Content Model were developed.

ICPC-3 FRAMEWORK

The new ICPC-3 content is based on a number of user needs:

- the need to capture person-centeredness in registrations in daily practice
- the need to support shared decision-making
- the need to support both comparability of data in a diversity of health care settings and exchange of data
- the need to support capturing of data for research and policy

The first step was to develop a framework for visualising person-centeredness in the components that the ICPC-3 should be built on. In the process of framework development, several schemes were discussed. The traditional biaxial structure used in the ICPC-2 has no flexibility to expand on, as it is meant for coding purposes and not to serve as a guiding framework. Recently, as part of a discussion on how to better depict person-centredness, alternative International Classification of Functioning, Disability and Health (ICF) schemes have been published.[1] One of these schemes

DOI:10.1201/9781003197157-2

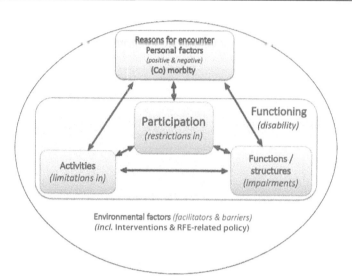

FIGURE 2.1 ICPC-3 Framework.

seemed suitable for the ICPC as well. The scheme has been adapted to capture the core components of the ICPC-3 (as shown in Figure 2.1).

Figure 2.1 is a visual representation of the ICPC-3 with the focus on Functioning and Environmental factors. Functioning (shown in green) is positioned in the centre of the Framework, with Participation as the core of functioning. Environmental factors (in blue) include Interventions and RFE-related policy as well as RFEs, Personal factors and (Co)morbidity (enclosed in violet and positioned at the top of the Framework to indicate their importance).

ICPC-3 CONTENT MODEL

The next step was to design the Content Model in line with the new ICPC-3 Framework.

In the third revision of the ICPC, the content is based on the ICPC-2 with additional categories to capture Functioning, Functioning related, Environmental factors, and Personal factors (as presented in Figure 2.2) and a restructuring and expanding of the categories for Prevention.

Also new are the Regional extensions for regional and national use; these are for the African, European and South American regions. These are based on the principle that additions need to be relevant for primary health care and supported by evidence of frequency on international or regional (national) level.

The Content Model is composed of two parts.

One part includes the **descriptive characteristics**:

- the name of the class and, within each class, the name of the categories and, if relevant, a textual description, what is included or excluded, index terms and synonyms, a coding hint and a place for a note or mark about the class

THE ICPC-3 CONTENT MODEL

Any Class/Category in ICPC is represented by:

Descriptive characteristics

1. **TITLE of Entity: Name of class**
   a. Textual description, concise and detailed
   b. Inclusion – Exclusion - Index terms/synonyms – Coding hint - Note
2. **Type of Entity**
   a. Non-problem related
      • Prevention
      • Screening
   b. Body/Organ System
      • Symptoms, complaints and abnormal findings
      • Diagnosis and Health Problems
   c. Social Problems (Z-chapter)
      • Social and Environment
   d. Interventions (patient related) and Processes (administrative)
   e. Functioning
      • Activity and Participation, Functions
   f. Functioning related factors
      • Personal factors, Environmental factors
   g. Regional Extensions
   h. Emergency Codes
3. **Extension Codes**
   a. Severity and/or – existing severity scales- ICF scale, stages
   b. Duration, course

Maintenance attributes

A. **Unique identifier**

B. **Attributes** (subset, adaptation, and special view flag) for:
   1. Classes – in disease component (congenital, infectious, neoplasm, injury, immunology, life-style, other, unknown)
   2. Classes – in environment component – context and contact Reason)
   3. Country adaptation
   4. Research
   5. Special indices (e.g. Primary Health Care Indicators, Public Health Care Indicators, and First aid or Resource Groupings, Case-mix)

C. **Hierarchical relationships**
   Parents and children in the ICPC structure:
   Chapter
   Component
   Classes/subclasses

D. **Reference relationships**
   References to classes as in ICPC-1, ICPC-2, ICD-10, ICD-11, ICF, ICHI, GBD, SDG's, UHC, and terms as in SNOMED CT etc.

E. **Other rules**

FIGURE 2.2 ICPC-3 Content Model.

- the type of entity, organised as: Non-problem related; Body/organ system, subdivided into Symptoms, complaints and abnormal findings, and Diagnoses and diseases; Social problems; Interventions and processes; Functioning, subdivided into Activity and participation, and Functions; Functioning related, subdivided into Environmental factors and Personal factors; Regional extensions; and Emergency codes
- if relevant, classes can also be expressed more meaningfully by adding severity scales, stages of processes, duration and course

The other part covers **attributes for maintenance**:

- a Unique identifier/class code
- meaningful Attributes (shown in different colours) for classes so as to be able to distinguish and organise classes in terms of symptoms, clinical findings and complaints and concerns in the component for Symptoms, complaints and abnormal findings
- in the component for Diagnoses and diseases, congenital, infectious, neoplasm, injury, immunology, lifestyle, other, and unknown diagnosis
- contact reason as an attribute for all classes in Chapter A1
- process as an attribute for all classes in the Interventions and processes chapter

- context as an attribute for classes in Chapter Z and in Component 2R0 from Chapter 2R Functioning Related
- personal factor as an attribute for all classes in Component 2R3 Personality functions
- in Regional extensions, the regional adaptation is subdivided into Regional chapter and Regional component (the colour attributes in this part are the same as the colour attributes in the core of the ICPC-3)
- hierarchical relationships (parents, children and grandchildren) in the ICPC structure: chapters, components and classes/subclasses
- reference relationships with links to the ICPC-1, the ICPC-2, the ICD-10, the ICD-11, the ICF, the International Classification of Health Intervention (ICHI), universal health coverage (UHC), and SNOMED CT
- attributes for Research and Special indices have been provided in the Content Model for future applications, but these are not used yet

Based on the Content Model, the division of chapters and components has been derived and the structure for the ICPC-3 has been built within an electronic authoring tool, the Classification Manager.

After development of the Framework and the Content Model, a review process was carried out. This was:

- based on a review of ICPC-2 content by primary health care experts
- informed by registration data from daily practice
- informed by participation of content experts within the WHO Family of International Classifications (WHO-FIC) working group on the ICD-11 for primary care

SELECTION OF CLASSES

The selection of classes, also called categories, is based on the classes in the previous versions of the ICPC and enriched by proposals from WICC and from the Consortium members themselves.

As it has been from the start, the selection of classes is based on frequency of occurrence in daily practice. The classes all have their own code. Less frequently used morbidity concepts are captured as inclusions within the main classes.

RELATIONS WITHIN INTERNATIONAL CLASSIFICATIONS AND SNOMED CT

An important feature of the ICPC is that it is part of the WHO-FIC. In this context, it fulfils a role as *the* RFE classification.

In Figure 2.3, the ICPC-3 Content Model is visualised in relation to the classifications that make up the WHO-FIC. These classifications are the ICD-11, the ICF and the ICHI, which at this moment are three separate entities.

In Figure 2.3, the Content Model refers to the two parts of the ICPC-3, as explained earlier. The part in the orange box shows the high-level categories of the ICPC-3; the grey area surrounding the orange box represents the underlying structure and attributes for maintenance of the ICPC-3 content (as shown in Figure 2.2).

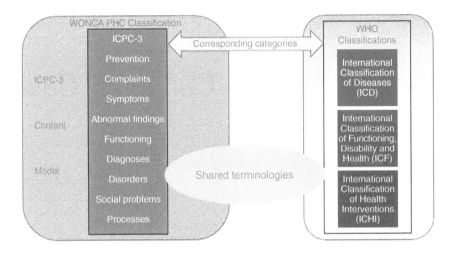

FIGURE 2.3 ICPC-3 in relation to the WHO Family of International Classifications.

The categories in the orange box are the parts in a single classification that cover all parts of the registration process in primary care in a structured and integrated manner.

The arrow at the top of the figure represents the principle of correspondence between the ICPC-3 categories (and classes) and the categories in the separate classifications of the WHO-FIC, in which medical terms, terms for functioning and terms for interventions represent the same concepts or meaning. Given the need to compare and exchange data between professionals and systems, this underlines the importance of having references in the ICPC-3 to the ICD, the ICF and the ICHI, but also to other terminologies such as SNOMED CT.

The ICPC-3 also contains specific primary health care terms not present in or not suitable for the WHO classifications and not present in SNOMED CT.

The overlapping oval indicates that the same terms from different sets of terminologies are used; for example, where international standards are available, the Foundational Model of Anatomy for anatomical entities.

The new ICPC-3 is already used by the WHO Primary Health Care department in the context of the UHC Compendium[2] as a structuring framework, and it is linked to WHO international classifications. This allows communication between the ICPC-3 and the other classifications and complementary usage. Ongoing cooperation between WONCA and the WHO Primary Health Care and WHO-FIC network exists for the harmonisation of the ICPC-3 with the UHC Compendium and the WHO, ICD-10, ICD-11, ICF and ICHI classifications.

REFERENCES

1. Heerkens YF, de Weerd M, Huber M, de Brouwer CPM, van der Veen S, Perenboom RJP, van Gool CH, ten Napel H, van Bon-Martens M, Stallingaand HA, van Meeteren NLU. Reconsideration of the scheme of the International Classification of Functioning, Disability and Health: incentives from the Netherlands for a global debate. *Disability and Rehabilitation* 2017; 40 (5): 603–611.

2. WHO. UHC Compendium, Health Interventions for Universal Health Coverage. www.who.int/universal-health-coverage/compendium

CHAPTER 3

The Primary Care Use Case

PRESENT CLASSIFICATIONS AND USE CASES

The classifications that are used today, in particular the ICD-10 and the ICPC-1 and ICPC-2, are not built to capture person-centredness, such as functions, activities, participation and the personal environment, in a structured and integrated manner and in one classification.[1,2,3,4] WHO and WONCA started a collaborative project in an attempt to derive a primary health care linearization from the ICD-11 and concluded that primary health care data used in daily practice cannot be captured within a classification driven prominently by disease.[5] The emphasis of the ICD is too much on hospital-related diseases and disorders, which requires a different classification. The ICPC-3 includes self-limiting diseases and health problems as presented in primary health care practice, which are lacking in the ICD; for example, shoulder syndrome, neck syndrome and back syndrome. The feasibility of deriving an ICD-11-based linearization for primary health care is still in the explorative phase.[6] The most important use cases that underline the need for a new classification for all of primary health care are presented in the box below.

Use Case 1: **Capture Person-Centred Registration in Daily Practice**
There is a desire to be able to broaden the scope of patient contacts and information from a pure medical perspective to a person-centred perspective. In primary health care, the patient is the centre of attention, but this is not reflected in the way data are collected and registered. In most patient cases, medical information is focused on medical diagnoses, and a lot of 'other' valuable information can only be found in notes or in the minds of health professionals. This includes the reasons why patients contact the health care system, how they perceive their health situation and what they expect in terms of results. With the high increase in the number of people ageing, and therefore an increase in comorbidity and possible functioning problems, a broader view of a person will be required to provide adequate care.

Use Case 2: **Shared Decision-Making**
In daily practice, there is a desire to involve the patient actively in the decision-making process, thus giving the patient a more prominent role in the provision of information, sharing this information and jointly setting goals for improved functioning, with a focus on participation = being involved in life situations.[7] This requires registration of patient-related data in a coherent manner.

DOI:10.1201/9781003197157-3

Use Case 3: Comparability; Diversity of Health Care Settings; Interchangeability
Primary health care takes place in many different settings and is provided by different professionals, all using their own sets of tools, terms and classifications to capture patient data. In most health care settings, the same data are registered repeatedly. There is no unified framework or language that could serve as a building block for the diagnostic and therapeutic process. This makes it very difficult to share and compare data.

Use Case 4: Research and Policy
Policymakers, funders and researchers need to have information about the epidemiology of their communities, and they need to understand what is happening within primary care to improve health services. More and more, the influential political bodies in the world, such as the Organisation for Economic Co-operation and Development (OECD), the United Nations and local governments, are showing interest in the well-being/functioning of the population.[8] Using only diagnoses to describe the health of the population has become too limited, especially in a world where people are more concerned with their participation in society and well-being. Too often and for too long, it seems that highly specialized care has been overvalued rather than there being a focus on what people want to receive in care and how their health situation is presented.

WHAT ARE THE MAIN CHANGES FROM THE ICPC-2?
Inclusive for All Primary Health Professionals

The ICPC-3 has been developed with all primary health care providers in mind, not only family doctors or general practitioners.

Inclusive for All WONCA World Regions

The ICPC-1 and ICPC-2 were mainly developed in industrialized countries, and as already mentioned in the ICPC-2-R, modified modules were necessary; for example, for tropical conditions. These classes have been added in the ICPC-3 in the Regional extensions chapter, for regional and national use. Currently, the Regional extensions chapter consists of classes from the African, European and South American regions. These are based on the principle that additions need to be relevant for primary health care and supported by evidence of frequency on an international or regional (national) level.

But it is not just that classes for tropical conditions have been added; new needs have also arisen:

- the need to focus on regions and countries
- the need to better code themes around prevention

- the need to record functioning of patients
- the need for some new classes, for which there was no space in the ICPC-2

Classes that had been incorrectly positioned in the ICPC-2 – such as infections and trauma of the skin in the symptoms/complaints part of the body system Skin – are now in the correct position within the classification. Additional categories for Functioning, Functioning related, Environmental factors and Personal factors are now part of the classification, and a restructuring and expansion of the categories for prevention has been completed.

What has stayed the same is that the granularity is still based on the frequency of what is presented worldwide in primary health care; though local frequencies are taken into account as well. And if more detail is required, the ICPC-3 opens up to other international classifications and a clinical terminology.

The biaxial structure of the ICPC-2 has been integrated in a dynamic and modern classification, supporting new technological requirements to be able to compare and exchange data between systems, a so-called 'interoperability' (see Figure 2.3), and backwards compatibility with previous versions of the ICPC. 'Dynamic' means that updates can and will be incorporated as required. Its prime use is online or on a desktop computer, laptop, tablet or mobile phone. To serve all users, a paper-based desk version is also available. For implementation in electronic health records, or for statistics purposes, several export formats are available, including crosswalks to the ICD-10, the ICD-11, the ICHI, the ICF, SNOMED-CT and several validated questionnaires.

The creation of more space and a different structure has led to a different coding scheme. Chapter 13 presents conversion tables from the ICPC-3 to the ICPC-2.7 and to the ICPC-1.

SUMMARY

Now, RFE, functioning (activities and participation) and personal preferences can be linked to morbidity. The ICPC-3 includes all those classes/concepts in primary care that can lead to better decisions by providers and policymakers. It includes the new approach to health – person-centredness – providing a professional language that is used in daily practice by primary care providers.

REFERENCES

1. International Statistical Classification of Diseases and Related Health Problems, Tenth Revision (ICD-10). Geneva, World Health Organization, 1992.
2. WONCA International Classification Committee. *International Classification for Primary Care, Second Edition (ICPC-2)*. Oxford: Oxford University Press, 1988.
3. Van den Muijsenbergh M, Van Weel C. The essential role of primary care professionals in achieving health for all. *Ann Fam Med* 2019; 17 (4): 293–295.
4. De Maeseneer J, Boeckxstaens P. James Mackenzie Lecture 2011: Multimorbidity, Goal-Oriented Care, and Equity. *Br J of Gen Pract* 2012; 62 (600): e522–e524.
5. Ten Napel H, Verbeke M, van der Haring E, Knupp D, Njoo K, Schrans D, Olagundoye O, Härkönen M, Letrilliart L, Frese T, van Boven K. *The ICPC-3 Development within the WHO-FIC Framework*. WHO-FIC Annual Meeting, Banff, Canada, 2019, Poster 602.

6. van Gool CH, Hanmer L, Hardiker N, Maart S, Madden R, Shin D, Sive W, Vikdal M, Virtanen M, Whitelaw L. *WHO-FIC Primary Health Care Linearization: A Report on Process and Progress.* WHO-FIC Annual Meeting, Banff, Canada, 2019, Poster 601.

7. Charles C, Gafnv A, Whelan T. Shared decision-making in the medical encounter: what does it mean? (or it takes at least two to tango). *Soc Sci Med* 1997; 44 (5): 681–692.

8. OECD. *Health at a Glance 2017: OECD Indicators.* OECD Publishing, Paris, 2017. http://dx.doi.org/10.1787/health_glance-2017-en

Description, Inclusion, Exclusion, Coding Hint, Note, Index Terms and Cross References

When reading this chapter, using the browser (https://browser.icpc-3.info/) is recommended. This gives a better insight into the structure of the classification.

Several rubrics are distinguished within each class: description, inclusion, exclusion, coding hint, note, index terms and cross references.

DESCRIPTION

The description is a short characterisation of the entity (class/component/chapter) that states things that are always true about that entity and necessary to understand the scope of the entity. The description should minimise variability in coding. Where possible and necessary, the classes have a description.

Examples

Description of Class **AS01 General pain in multiple sites**:
    Pain is an unpleasant sensory and emotional experience associated with actual or potential tissue damage, or described in terms of such damage. Often, pain serves as a symptom warning of a medical condition or injury. In these cases, treatment of the underlying medical condition is crucial and may resolve the pain. However, pain may persist despite successful management of the condition that initially caused it, or because the underlying medical condition cannot be treated successfully. Chronic pain is pain that persists or recurs for longer than 3 months.

Description of Class **FD01 Infectious conjunctivitis**:
    Presumed or proven infectious inflammation of conjunctiva.

INCLUSION

Within the classes, there are typically other optional entities. These entities are known as 'inclusions', and they are given, in addition to the title, as examples of issues to be classified to a particular class. They may refer to different conditions. They are as such not intended as a subclass of the category, but can be used within the core classification

DOI:10.1201/9781003197157-4

and can have a code. The goal of inclusions is to inform the health provider of what falls within classes. The lists of inclusion terms are by no means exhaustive. In principle, inclusions contain terms for conditions that are less frequent.

Example

Inclusion for Class **FD01 Infectious conjunctivitis**:

> bacterial conjunctivitis [with the regional code] FD01.00
> conjunctivitis NOS [not otherwise specified]
> viral conjunctivitis FD01.01

Regional extension codes are explained in Chapter 10.

EXCLUSION

Certain classes have exclusions, a list of conditions which are classified elsewhere. Exclusions serve to guide the user to the relevant code in the classification and as a cross reference in the ICPC to help to delimit the boundaries of a class.

Example

Exclusion of Class **FD01 Infectious conjunctivitis**:

> allergic conjunctivitis with/without rhinorrhoea FD65
> flash burn FD37
> other eye inflammation or eye infection FD03
> trachoma, chlamydia conjunctivitis FD04

CODING HINT AND NOTE

The ICPC-3 makes limited use of coding hints and notes.

   A **coding hint** refers to another class that may better reflect what the coder is looking for.

Example

Coding hint for Class **AD03 Rubella**:

> rash generalised SS06
> viral exanthems AD13

A **note** serves in most cases as a directive on how the class should be used.

Examples

Chapter Z, Social problems

In all classes in Chapter Z, it is indicated that the diagnosis of problems requires the patient's agreement on the existence of the problem and desire for help.

Note for Class **ZC01 Partner relationship problem**:

The diagnosis of problems in the relationship between family partners requires the patient's agreement on the existence of the problem and desire for help.

Chapter P, Psychological, mental and neurodevelopmental

In classes related to substance abuse – PS12, PS14, PS15 and PS16 – versions of the following are included as notes:

> [The class] should take into account the considerable differences between countries and cultures. A doctor can decide to label an episode as '[class]' without the patient's agreement, and consequently also without the patient's willingness to any medical intervention.

INDEX TERMS

Index terms are listed primarily as a guide to the content of the class; they are in addition to the complete description or to illustrate it. The index terms list is selected from ICD-10, ICD-11, ICF, ICHI and SNOMED CT terms.

The search terms, used for indexing the complete ICPC-3, are based on the preferred term (the label of the class), the inclusion terms and the index terms.

The lists of index terms are by no means exhaustive. Synonyms and lay terms can also be included.

In addition, the synonyms used in the ICD-11, SNOMED CT, etc. are sometimes added to the list of index terms. There are two exceptions to this rule:

1. the words in the exclusion do not belong to the meaning of the class
2. where the references to the ICD, the ICF, SNOMED CT, etc. are too extensive, because there is no exact corresponding class/term available

The index terms are a necessity to be able to build a complete search index for the ICPC-3. This will be used for building the ICPC-3 thesaurus and can serve as a standalone version of the alphabetical index if so required.

Example

Index terms for class **FD01 Infectious conjunctivitis**:

blepharoconjunctivitis
chlamydial conjunctivitis
chronic conjunctivitis
follicular conjunctivitis
Herpes zoster conjunctivitis
mucopurulent conjunctivitis
parasitic conjunctivitis
purulent conjunctivitis

CROSS REFERENCES (LINKAGES)

The ICPC-3 content contains linkages to several standardised classifications, such as the ICD-10, the ICD-11, the ICF, the ICHI, the Diagnostic and Statistical Manual of Mental Disorders, 5th Edition, and clinical terminologies such as SNOMED CT, but also to previous versions of the ICPC-1 and the ICPC-2.7 and, where relevant, to the United Nations Sustainable Development Goals.

The linkages to these classifications and terminologies serve as a pathway from and to the ICPC. This is what is generally called a 'telescopic' or 'periscopic' view. Starting with the categories or classes in the ICPC-3, when more detail is needed, it is possible to zoom in to the ICD for diagnostic classes, the ICF for functioning or the ICHI for interventions. The other way around, when detailed data is received, it is possible to zoom out to the relevant ICPC-3 categories/classes.

With these linkages, the ICPC supports the principle of continuity of data within and between health care providers, but it also supports the use of the ICPC, or the ICD within a country, without losing the possibility to collect or exchange information for different purposes, such as direct patient care, research, reimbursement, aggregation of data, disaggregation of data, etc.

For the information exchange process, standardisation is required on a different level. Here, it is necessary to capture the meaning of the content using the same (clinical) terminologies; for example, the use of the Foundational Model of Anatomy throughout all related classifications and clinical terminologies.

In the ICPC-3 online browser, the codes or identification numbers (IDs) behind the terms in the ICD-10, the ICD-11 and SNOMED CT guide the user to the same code within the external classification browsers at WHO and SNOMED CT.

Example

Cross references for Class **FD01 Infectious conjunctivitis**:

- *ICPC-1* F70
- *ICPC-2* F70
- *ICD-10* no exact corresponding class 00
  - Blepharoconjunctivitis H10.5
  - Chlamydial conjunctivitis A74.0
  - Mucopurulent conjunctivitis H10.0
  - Other acute conjunctivitis H10.2
  - Viral conjunctivitis B30
- *ICD-11* no exact corresponding class 00
  - Blepharoconjunctivitis 9A60.4
  - Chlamydial conjunctivitis 1C20
  - Follicular conjunctivitis 9A60.1
  - Mucopurulent conjunctivitis 9A60.3
  - Viral conjunctivitis 1D84
- *SNOMED CT* infective conjunctivitis ID 299699004
  - bacterial conjunctivitis ID 128350005
  - chlamydial conjunctivitis ID 231861005

- chronic conjunctivitis ID 73762008
- follicular conjunctivitis ID 86402005
- herpes zoster conjunctivitis ID 410509003
- parasitic conjunctivitis ID 13816006
- purulent conjunctivitis ID 243321006
- viral conjunctivitis ID 45261009

ICD REFERENCES

The references to the ICD-10 in the ICPC-3 are based on the ICPC-2 to ICD-10 conversion, meaning that in some cases an exact corresponding class is available but, in addition, other ICD-10 classes are related. In some cases, there is no exact corresponding class, but some 'alike' classes are available. The selection of ICD-11 references is based on existing ICD-10 references as presented in the ICD-11. A high number of class names are unchanged from the ICD-10 to the ICD-11; and in the ICD-11, where more detailed classes have been introduced, an extensive browser search is offered to find similar classes.

SNOMED CT REFERENCES

The selection for SNOMED CT references is based on:

1. the existing cross references from SNOMED CT to the ICPC-2, developed and agreed in a collaboration between SNOMED International and WONCA International Classification Committee
2. frequency of the used search terms/concepts from the Dutch thesaurus with linkages between the ICPC-2 and the ICD-10
3. frequency of the search terms/concepts of the Belgium thesaurus with linkages between the ICPC-2 and the ICD-10
4. the index terms of the ICPC-3, regarding the ICD

CHAPTER 5

Episodes of Care
A Central Concept in Primary Health Care

Changes in the need for and use of classifications in primary care have continued since the last publication of the ICPC-2-R in 2005. Then, the main purposes of the classification was seen to be its use in registration data for daily practice by the health care provider, research, and policy formulation. However, its use has widened as research data and practical experiences with the ICPC, as well as the emergence of new concepts in general/family medicine, have resulted in new applications. The most important new applications of the use of the ICPC are:

- describing the construct of care episodes with the ability to relate care episodes to functioning and to have problems in functioning as episodes of care (EoCs). This is very important in order not to fall into the trap of omitting context by not analyzing the outcome of policy at a personal level.
- the use of the ICPC-3 as a dynamic and modern classification, supporting new technological requirements to be able to compare and exchange data between systems, a so-called 'interoperability'. 'Dynamic' means that updates can and will be incorporated as required. Its prime use is online and on a desktop computer, laptop, tablet or mobile phone. To serve all users, a paper-based desk version is also available. For implementation in electronic health records, or statistics purposes, several export formats are available, including crosswalks to the ICD-10, the ICD-11, the ICHI, the ICF, SNOMED CT and several validated questionnaires.

These new applications are closely related and depend on the use of the ICPC as the ordering principle of patient data gathered in primary health care. WHO (Alma Ata 1978) defined primary health care as follows:

> Primary health care is essential health care made universally accessible to individuals and families in the community by means acceptable to them, through their full participation and at a cost that the community and country can afford. It forms an integral part of the country's health care system, of which it is the nucleus, and of the overall socio-economic development of the community.

DOI:10.1201/9781003197157-5

Primary health care can be delivered by primary health care nurses, physicians or health professionals with a shorter medical training ('barefoot doctors', physician assistants). This is quite like WHO and UNICEF's 2021 definition of primary health care:

> a whole-of-society approach to health that aims at ensuring the highest possible level of health and well-being and their equitable distribution by focusing on people's needs and as early as possible along the continuum from health promotion and disease prevention to treatment, rehabilitation and palliative care, and as close as feasible to people's everyday environment.[1]

The WONCA definition of general/family practice refers to 'a physician who provides personal, primary and continuing comprehensive health care to individuals and families'.[2]

EPISODE OF CARE

The EoC allows for grouping of data over time. Health care providers can use these data to improve continuity and coordination of care.[3] The ability to collect data using the EoC also creates more insight into the processes related to certain conditions over time and, thus, to a greater understanding of what is needed and the costs associated with it. EoCs are distinguished from episodes of illness or disease in a population. An EoC refers to a health problem or disease from its first presentation to a health care provider until the completion of the last encounter for that same health problem or disease (Figure 5.1).

RFEs, functioning, health problems, diagnoses, and process of care and interventions shape the core of an EoC consisting of one or more encounters, including changes in their relations over time ('transitions'). An EoC, consequently, refers to all care

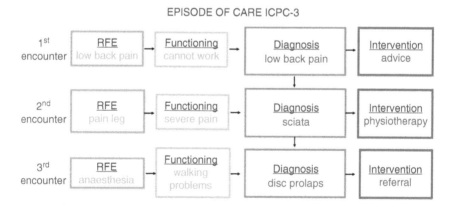

FIGURE 5.1 Episodes of care.

provided for a discerned health problem or disease in a particular patient. The 'large majority of personal health care needs', the 'comprehensiveness', the degree of 'integration', 'accessibility' and 'accountability' can be assessed when EoCs are classified with the ICPC in a computer-based patient record. Also, the concept of RFE proved to be an innovative and practical operationalisation of the patient's perspective and demand for care; the validity of the RFE – as coded by family doctors when compared with the patient's point of view after the encounter – is consistently very high.[4]

REASON FOR ENCOUNTER

The RFE connects care providers with the client.[3] That is why it is so important. The RFE contains everything a person seeking help has internalised, her or his personal environment, their past and their views. The RFE enables recording of the problem as expressed by the person, followed by coding of presented problems in terms of symptoms or complaints, limitations in activities of barriers in participation, but also requests (for prescriptions, referrals or investigations) and cognitions, emotions, worries, concerns or fears that bring people to contact health services.

The RFE has been established to be a practical source of patient information, also useful for research and education. This is illustrated by epidemiological data from the FaMe-net project (Family Medicine network, a merger of former the Transition Project and Continuous Morbidity Registration Nijmegen) in standard format (www.famenet.nl/).

Beginning with the RFE allows the determination of the probabilities of any given health problem at the start or during follow-up of the episode per standard sex age group. Therefore, the top 10 problems related to fever at the start of an episode show clinically important differences between children aged 5–14 and adults aged 65 and over (Table 5.1).

The reverse procedure is equally relevant from a clinical point of view: what RFEs were presented at the start of a problem in each standard sex age group? This is given in Table 5.2 for pneumonia.

HEALTH PROBLEM/DIAGNOSIS IN THE CARE EPISODE

The health problem or diagnosis is central to the EoC. Many health problems are in fact medical diagnoses, but in primary care there are many other conditions such as fear of disease, symptoms and complaints not attributed to a disease (symptom diagnosis), or limitations (in activities) and barriers (in participation). Sometimes there is no apparent health problem involved in an EoC; for example, when it relates to a need for immunisation or screening, family planning, patient preferences or case finding. These contacts can also be related to first contact or request for certification. The ICPC includes all of these. The health problem may be qualified in terms of its status in the encounter and the certainty which the provider assigns to its diagnosis, and by using the extension codes to give more detail or meaning to the health problem. The status of the episode in an encounter can be specified as new to both health professional and patient, new to the health professional but previously treated outside the current provider system, or neither in the case of follow-up. In any environment, electronic or paper based, this can be easily

TABLE 5.1 Top 10 episode titles starting with fever (AS03) as the reason for encounter (prior probabilities)

Children 5–14 years old			ICPC-3 code
	N	%	
Upper respiratory infection	800	17.0	RD02
Fever	689	14.6	AS03
Other viral diseases NOS	441	9.4	AD14
Acute otitis media/myringitis	384	8.1	HD02
Tonsillitis acute	352	7.5	RD04
Influenza (proven) without pneumonia	310	6.6	RD07
Pneumonia	304	6.4	RD09
Acute bronchitis/bronchiolitis	251	5.3	RD06
Presumed gastrointestinal infection	159	3.4	DD05
Symptom/complaint throat	91	1.9	RS12
Total top 10	3,781	80.2	
Total	4,716	100.0	

Men and women aged 65+			
	N	%	
Pneumonia	572	17.9	RD09
Fever	440	13.7	AS03
Acute bronchitis/bronchiolitis	393	12.3	RD06
Cystitis/other urine infect NOS	288	9.0	UD02
Influenza (proven) without pneumonia	184	5.7	RD07
Upper respiratory infection	166	5.2	RD02
Other viral diseases NOS	113	3.5	AD14
Emphysema/COPD	95	3.0	RD68
Pyelonephritis/pyelitis acute	82	2.6	UD01
Sinusitis acute/chronic	77	2.4	RD03
Total top 10	2,410	75.3	
Total	3,202	100.0	

solved using 'flags'; for instance, in case of a known patient already diagnosed with diabetes, using an (X), a known patient with a new diagnosis of diabetes, using an (N), and a follow-up contact within the existing episode of diabetes using an (O) (see Figure 5.2, box G). Another aspect of an EoC is the extent to which the health professional is certain that his or her diagnosis is correct; this can be graded from uncertain to certain,

TABLE 5.2 Top 10 reasons for encounter in an episode of pneumonia (RD09)

Children aged 5–14	N	%	ICPC-3 code
Cough	297	40.7	RS07
Fever	233	32.0	AS03
Shortness of breath/dyspnoea	39	5.3	RS02
Pneumonia	23	3.2	RD09
Med. examination/health evaluation/ partial	15	2.1	R102
General weakness/tiredness	15	2.1	AS04
Vomiting	10	1.4	DS10
Upper respiratory infection	9	1.2	RD02
Generalised abdominal pain/cramps	9	1.2	DS01
Ear pain/earache	7	1.0	HS01
Total top 10	657	90.2	
Total	729	100.0	

Men and women aged 65+	N	%	
Cough	722	25.5	RS07
Shortness of breath/dyspnoea	474	16.7	RS02
Fever	425	15.0	AS03
Pneumonia	147	5.2	RD09
General weakness/tiredness	125	4.4	AS04
Med. examination/health evaluation/ partial	80	2.8	R102
General deterioration	67	2.4	AS06
Provider-initiated episode new/ongoing	54	1.9	R501
Administrative procedure	52	1.8	R601
Medication/prescription/injection	49	1.7	R201
Total top 10	2,195	77.4	
Total	2,834	100.0	

but a standard recording of this grading has not yet been agreed on. The description and inclusion criteria for use of classes in the ICPC-3 will, however, help to ensure that the label chosen for the episode is used consistently by all providers. The qualification of an EoC using the extension codes is discussed in Chapter 10.

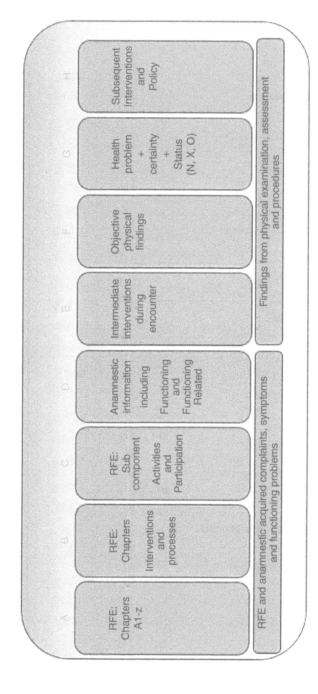

FIGURE 5.2 Structure for describing encounters.

INTERVENTIONS: THE PROCESS OF CARE

The specificity of the three-/four-digit ICPC process code to classify interventions is usually adequate for primary care practice. To give more specificity to the process code, the use of the codes in chapters A–Z as a first prefix is recommended. However, when drugs are prescribed, a drug code is needed. Because of the vast number of medications involved, and the idiosyncrasies of national drug availability, no internationally suitable code has yet been produced. The advice is to link the Arrêts de Travail en médecine générale à partir de la Classification Internationale du Fonctionnement, du handicap et de la santé (ATC) coding to the process classes -201 Pharmacotherapy and prescription and -202 Preventive immunisation and medication.

FUNCTIONING AND FUNCTIONING RELATED

With these classes, it is possible to describe Functioning and Functioning related aspects of the EoC. The classes acquire meaning when the patient makes a statement about them; for example, if the patient expresses a severe limitation in activity or barrier in participation, or if the patient expresses a level of problem with his/her energy, memory or balance.

ICPC-3 AND PATIENT RECORDS

The core of a computer-based patient record is data coded with the ICPC, which is language independent: this enhances the use of practice records for a comparison of data from different countries, and it supports the development of general/family practice as an internationally well-developed profession with a well-defined and empirically based frame of reference. The availability of the ICPC-2 in 19 languages and the growing number of translations of the ICD-10 accompanied by alphabetical indexes allow family doctors in many countries to incorporate a detailed language-specific thesaurus in their system, at the same time using the ICPC to systematically structure their records and the database in a more standardised way. The same developments are envisaged for translations of the ICPC-3.

An electronic patient record can be helpful in properly registering the data in an EoC. The system can warn the provider when she or he tries to enter a follow-up encounter for an episode that has not yet been established in the database, or whenever a new one is started even though an episode with the same title already exists. This is, obviously, vital to ensure the quality of daily recording. Pop-up screens can be used to display options at the time of coding in computer-based records and a good data system will be able to display these interrelationships between multiple health problems and provide data on comorbidity.

FURTHER DEVELOPMENTS

The original three basic elements of encounters to be coded with the ICPC (RFE, health problem and interventions) enriched with a fourth element, Functioning and Functioning related, have now been extended with eight data entry options (A–H) for computer-based patient records (Figure 5.2). The RFE is recorded in three sections: patient symptoms and complaints; patient requests for interventions;

and patient-expressed limitations in activities and barriers in participation (A–C). RFEs in the form of symptoms, complaints or health problems/diagnoses should be distinguished explicitly from those in the form of requests for interventions such as a prescription, an X-ray, a referral or advice and those in the form of limitations in activities or barriers in participation. Why is this important? Requests for a certain intervention are often followed by this intervention being performed – when patients ask for medication or a blood test, they often receive it. Since patients do actively influence the care provided by health providers, it is important to explicitly document this.

It seems useful for the future to also record the clinical anamnestic findings, including functions and functioning-related information (D), separate from the RFE. All relevant classes can be used for this purpose. It should be noted that the ICPC-3 does not yet include a classification of objective physical findings by the health provider (F).

Both new applications, coding of anamnestic data and objective physical findings, could be included in the encounter and episode structure of a computer-based patient record.

The use of RFEs and anamnestic data to estimate prior probabilities is clearly very useful. The difference between a symptom expressed by the patient as an RFE or elicited by the physician is retained, and the probabilities can be calculated separately if required.

Processes of care are recorded as immediate (those occurring during the encounter, E) or subsequent (interventions after the diagnosis or problem has been identified, H). The difference between what is in fact being done by the health provider at the time of the encounter and what is expected to follow is important for the analysis of utilisation data, inter-health care provider variation and application of guidelines. It also allows better understanding of the shift from prior probabilities in the first encounter of an EoC to the later probabilities during follow-up. For recording more specificity in interventions, not provided in the ICPC-3, a more specific process classification can be used in addition and linked to the ICPC. This could be ATC for -201 and -202, LOINC (Logical Observation Identifiers Names and Codes database) or a locally used list of laboratory tests, etc. It is not feasible to include this level of detail in the ICPC-3. Development of these relations is an ongoing activity.

REFERENCES

1.   WHO and UNICEF. *A Vision for Primary Health Care in the 21st Century: Towards Universal Health Care and the Sustainable Development Goals.* 2018. Geneva, WHO and UNICEF.
2.   Bentzen N (ed.). An international glossary for general/family practice. *Fam Pract* 1995; 12: 341–369.
3.   Lamberts H, Hofmans-Okkes IM. Episode of care: a core concept in family practice. *J Fam Pract* 1996; 42: 161–177.
4.   Hofmans-Okkes IM. An international study into the concept and validity of the reason for encounter. In: Lamberts H, Wood M, Hofmans-Okkes IM, (eds.), *The International Classification of Primary Care in the European Community.* Oxford: Oxford University Press, 1993, 34–44.

CHAPTER 6

Standard for Use of Reason for Encounter

The ways of coding information using the ICPC vary somewhat according to the type of information being recorded; for example, RFE, health problem or intervention. To promote consistent recording and therefore better comparability of data between centres, the following standards are suggested.

REASON FOR ENCOUNTER

Patients normally start the consultation with a spontaneous verbal statement on why they are visiting the health professional, known as the RFE. It is the beginning of the interaction and precedes interpretation by GPs and patients. The RFE is the literal expression of the reason(s) why a person enters the consultation room, translated into an ICPC code by the health provider. It represents that person's need for care. The RFE can be presented in the form of symptoms and complaints ('abdominal pain, a rash') but also as self-diagnosed diseases ('I've got the flu'), a problem carrying out an activity ('I cannot work') or requests for a particular intervention.

The primary care provider should identify and clarify the RFE as stated by the patient without making any judgments as to the correctness or accuracy of the reason. The patient statement is translated into a classification term and coded. This use of the classification is guided by three principles:

1. The RFE should be understood and agreed on between the patient and the provider and should be recognised by the patient as an acceptable description.
2. The ICPC class chosen should be as close as possible to the original statement of the reason given by the patient and must represent a minimal or no transformation by the provider. However, clarification of the patient's RFE within the framework of the ICPC is necessary so that the most appropriate class can be applied.
3. The description and inclusion criteria listed for classes, for use in recording health, are NOT to be used, since the RFE is to be documented from the patient's point of view, based entirely on the patient's statement of the reason.

Almost all parts of the classification are applicable, as patients can describe their reasons for seeking health care in the form of symptoms or complaints, as requests for services, as activities and participation problems or as health problems.

The way in which a patient expresses his/her RFEs determines which chapter and which (sub)component to use, except the classes that fall under the Functioning related component and the Functions subcomponent. These classes can be used by

DOI:10.1201/9781003197157-6

the provider to further explore the RFE and EoC for similar concepts in Symptoms and complaints (see Figure 5.2).

CHOOSING THE CHAPTER CODE

To code the RFE, it is necessary to first select the appropriate chapter, assign the correct one- or two-digit alpha code, and then assign the two- or three-digit numeric code in the relevant (sub)component, such as a symptom or complaint, a diagnosis, limitations in activities and participations, or an intervention. The search terms in the online ICPC-3 should be used when there is uncertainty about the chapter or (sub) component in which a specific RFE should be placed.

Chapter A1 is used for RFEs that relate to need for immunisation or screening, family planning, patient preferences or case finding. Chapter A is used for RFEs that relate to unspecified or multiple body systems, chapters B–W for RFEs related to body systems, and Chapter Z for RFEs related to social problems. RFEs related to processes of care are found in Chapter I, and RFEs related to limitations in activities and participation are in Chapter II.

When the ICPC is used for recording RFEs, five rules apply for the use of chapters, and there are two rules specific to Chapter I Interventions and processes. Those rules are listed below with examples for the application of those rules.

Rule 1
Whenever the patient makes a specific statement, use his/her terminology.

Example

Jaundice, in the form of a diagnostic descriptive term, can be found in Chapter D (on the digestive system), but the patient may present this symptom as a yellow discoloration of the skin (Chapter S). If the patient expresses the problem as 'jaundice', the ICPC code is DS13 Jaundice. If, however, the patient states 'my skin has gone yellow', the correct code would be SS07 Skin colour change, even though the health care provider is positive that the diagnosis is some form of hepatitis.

Rule 2
The RFE should be coded as specifically as possible and may require some clarification by the provider.

Example

Chest pain can be coded as AS12 Chest pain, or as KS01 Pain, pressure, tightness of heart, or as RS01 Pain respiratory system, or as LS04 Musculoskeletal chest symptom or complaint. The decision as to the correct selection is not based on the opinion of the provider as to the type of chest pain but, rather, to the way the patient expresses his/her RFE when clarification is sought by the provider.

'It's all over my chest ...' AS12
'My chest hurts when I cough' RS01
'I have chest pain ... I think it's my heart' KS01
'I have chest pain after falling down stairs' LS04

Rule 3
When the patient is unable to describe his/her complaint, the reason given by an accompanying person (e.g., a mother bringing in a child or relatives accompanying an unconscious patient) is accepted as being the reason stated by the patient.

Rule 4
If the patient indicates a limitation in activities or a barrier in participation, the degree of limitation must also be assessed using the problem scale value.

Rule 5
Any problem whatsoever presented verbally by the patient should be recorded as an RFE. Multiple coding is required if the patient gives more than one reason. Code every reason presented, at whatever stage in the encounter it occurs.

Example
'I need my asthma tablets. Also, my knee hurts' – R201, LS14.

If afterwards the patients asks, 'What is this lump on my skin?' or says, 'I can't climb stairs', those are also coded as RFEs – SS04 and 2F28 PV.3

CHOOSING THE (SUB)COMPONENT FROM THE CHAPTERS
Symptoms and Complaints in Chapters A–Z

The most common reasons patients report for seeking health care are presented in the form of symptoms and complaints. This implies that the Symptoms and complaints component of chapters A–Z will be used extensively. These symptoms are specific for each chapter; nausea (DS09) is found in the Digestive System chapter, while sneezing (RS09) is in the Respiratory System chapter. While most of the entries in this component are symptoms specific to the chapter in which they are found, some standardisation has been introduced for ease of coding.

STANDARDISATION OF CLASSES IN SYMPTOMS
AND COMPLAINTS IN CHAPTERS A–W

Throughout most of the chapters, except for Chapter A1, Chapter Z, Chapter I and Chapter II, the content within the -S component is organised as follows:

- -S01 to -S49 Symptoms and complaints
- -S50 to -S89 Abnormal results and physical findings
- -S90 to -S98 Concern or fear a disease or condition (cancer included)
- -S99 Other specified symptoms, complaints, or abnormal findings

The first class in every chapter relates to the symptom pain. Examples of these are ear pain or ache (HS01) and headache (NS01).

Code -S50, and sometimes also a few others, is used when the patient indicates an abnormal physical finding in themselves. Examples are:

'I think my blood pressure is low' KS50
'I have underweight' TS50

Code -S90, and sometimes also a few others, is used when the patient expresses concern about or fear of cancer or some other condition or disease. Examples are:

'I'm afraid I have TB' AS90
'I'm worried that I have cancer of the breast' GS93
'I'm scared of venereal disease' GS92

Even if the provider thinks that such an expressed fear is unwarranted or not logical, it constitutes the patient's RFE.

In each chapter, the component and subcomponent code -99 is the residual or 'ragbag' class for that (sub)component. This contains uncommon and unusual classes which do not have a separate class or are not part of the inclusion terms of other classes, and it can also be used for classes which are not clearly stated. The class 'not specified' is avoided, because in all cases it is necessary to be as specific as possible. At all times, the index terms should be consulted for synonymous terms in other classes before using this class.

Limitations in Activities and Barriers in Participation, Subcomponent 2F0 from Chapter II

Classes 2F01–2F69 should be used when the patient's RFE is expressed in terms of limitations or barriers which affect activities and participation in daily life and social functions. Always use the problem scale value.

Examples
'I cannot climb stairs because of the cast they have put on my leg for my fractured ankle' 2F28 and PSV.3 and LD36 (Component D, Diagnoses and diseases)

'I can't work in the office because I can't sit for any length of time because of my haemorrhoids' 2F58 PSV.3 and 2F21 PSV.3 and DD84 (Component D, Diagnoses and diseases)

Components of Interventions and Processes (Diagnostic, Therapeutic and Preventive Interventions, Programmes Related to Reported Conditions, Test Results, Referrals, and Administrative)

The reasons included in this concept are those in which the patient:

a. seeks some sort of procedure, such as 'I'm here to have a blood test' (-105)

Further clarification by the provider is often necessary to identify the most appropriate chapter code.

Examples
The patient may request a particular procedure in connection with an expressed problem or as a single demand, such as:

'I want the doctor to examine my heart' K102
'I think I need to have my urine tested' (-106)
'I need a vaccination' (-202)

To select the appropriate alpha code, clarification by the provider is necessary to find out why the patient thinks he or she needs a urine test. If it is because of a possible bladder infection, the code is U106; if it is because of diabetes, the code is T106. If the result of an X-ray which is being requested refers to a barium meal, the code is D401. The code for a request for vaccination against rubella is A202.

b. requests a treatment or when the patient refers to the physician's instructions to return for specific treatment, procedure or medication as the RFE

Further clarification by the provider is often necessary to identify the most appropriate code.

Examples
'I need my medication' (-201). If the patient expresses the reason why he is taking the medication or the provider knows the reason, select the appropriate alpha code; for example, for a sinus infection, the code would be R201.
'I'm here to have my cast removed' (-207). If it is evident that, for instance, the patient had a fracture of the left arm, the correct alpha code would be L.
'I was told to come for removal of the stitches today' (-207). Although, at first, one might assume that all suture removal would be in the chapter on skin, the patient might have stitches from eyelid surgery F207 or from a phimosis operation G207.

c. may request a care programme – a care programme consists of a combination of various interventions related to a reported condition

Example
'I've come for my diabetes programme' T308

d. is specifically requesting the results of tests previously carried out

This subcomponent should be used when the patient is specifically requesting the results of tests previously carried out. The fact that the results of the test may be negative does not affect the use of this component. Often the patient will request the test result and want to know about its consequences and seek more information on the underlying problem. In that case, also consider using the additional code -203 (Health education, advice and diet).

Examples
'I've come for the result of an X-ray of my ankle' L401

'I need the results of my blood test'. If the test was for anaemia, use code B401, if it was for hypercholesterolemia, use T401, and if the patient cannot specify, use A401.

'I am supposed to pick up the result of my urine test and take it to the urologist. I also want to know what he will do and which examinations and treatment I can expect' U401, U203

'I want to know the test results done by the specialist' -402. The class -402 should be used when the patient asks the result of an examination or test from another provider.

e. the RFE is to be referred to another provider

If the patient's RFE is to be referred to another provider, then -505 Referral to another primary care provider, -506 Referral to specialist, clinic or hospital, or -599 Other specified consultations, referrals and reasons for encounter can be used for this purpose. If the patient states his/her RFE is 'being sent by someone else', use -502.

f. the RFE for a problem initiated by the provider

When a provider initiates an episode or takes the initiative for the follow-up of an already existing episode of a health problem such as hypertension, obesity, alcoholism or a smoking habit, it is appropriate to code the RFE as -501 Encounter or problem initiated by provider. If the provider has advised the patient to come back for a control visit, this code is not used. Often the use of -102 Partial examination or health evaluation is the appropriate code.

Examples
A patient presents with a blocked ear due to earwax, which is removed. The provider measures his blood pressure (not an RFE mentioned by the patient) and finds it to be high, and the patient also receives advice about smoking. The patient's RFE and the related problems and treatment would be recorded as follows:

HS06 Plugged feeling in ear, HD66 Excessive ear wax, H204 (removal of earwax) K501 (provider initiated), KS51 Elevated blood pressure, K102 (checking of blood pressure)

P501 (provider initiated), PS14 Tobacco smoking problem, P203 (advice to stop smoking)

g. **administrative RFEs with the health care system include things such as examinations required by a third party (someone other than the patient), insurance forms which require completion, and discussions regarding the transfer of records**

Examples

'I need this medical insurance form completed' (A601)
'My fracture has healed, and I need a certificate to go back to work' (L601)

Diagnosis and Problems in Chapters A–W

Only when the patient expresses the RFE as a specific diagnosis or disease should it be coded in Component D in chapters A–W.

The RFE for a patient who is known to have diabetes but comes in complaining of weakness should not be coded as diabetes but as the problem expressed: AS04 General weakness or tiredness. However, if the patient states that he has come about his diabetes, the diagnosis 'diabetes' should be coded as the RFE (TD71 Type 1 diabetes mellitus or TD72 Type 2 diabetes mellitus).

If the patient names an RFE in the form of a diagnosis which the provider knows is not correct, the patient's 'wrong' RFE is coded rather than the physicians' 'correct' one; for example, a patient presenting with 'migraine' as the RFE when the provider knows it is a tension headache, or a patient who is known to have nasal polyps presenting with 'hay fever'.

Examples

'I am here because of my hypertension' KD73
'I come every month for arthritis in my hip' LD78

GENERAL RULE

Rule

Classes from more than one component, or more than one class from the same component, can be used for the same encounter if more than one reason is presented by the patient.

Example

'I've had abdominal pain since last night and I vomited several times' DS01, DS10
'I have some abdominal pain and I think that I may have appendicitis' DS06, DD72

CHAPTER 7

Standard for Use of Health Problems and Non-Disease-Related Care Episodes

HEALTH PROBLEMS

After anamnesis and physical examination, the health provider makes a diagnosis/assessment that indicates the care episode in which the encounter takes place. The diagnosis/assessment is the health provider's point of view. The episode label can be a symptom, a disease or problem, a problem in activity or participation, or a non-disease-related care episode such as visits related to a need for immunisation, to special screening examination and to public health promotion. The episode title can never be a Process, Intervention, Function, or Function related class.

To improve reliability of coding health problems using ICPC-3, almost all the classes have additional information classes to guide their use: descriptions, inclusion and exclusion terms, index terms and sometimes coding hints and notes. These are explained in Chapter 4.

GENERAL RULES FOR CODING HEALTH PROBLEMS AND NON-DISEASE-RELATED CARE EPISODES

Users are encouraged to register, during each encounter, the full spectrum of problems and care episodes managed in this encounter, including organic, psychological and social health problems and problems in activities and participation in the form of episode(s) of care (EoCs). Registering should be at the highest level of diagnostic refinement the user can be confident about, and should meet the description or inclusion for that class. In any data system, it is necessary to have clear and specific criteria for the way in which health problems or EoCs are registered. This applies particularly to the relationship between the underlying condition and manifestations when both may be available as classes in the classification. This is best illustrated by an example: A patient with ischemic heart disease may also have atrial fibrillation and resulting anxiety. It should be policy to include these as separate EoC manifestations which require different management. In this example, the atrial fibrillation and anxiety would be recorded as additional EoCs.

Some electronic systems accept that problems are coded with an intervention/process code. **This is not recommended or correct**. Interventions always take place in an

DOI:10.1201/9781003197157-7

EoC, and as indicated earlier, the care episode can, for example, relate to the need for immunisation or screening. Interventions carried out in these EoCs should be coded with the intervention codes in Chapter I, Interventions and processes, not with the classes in Chapter A1.

In ICPC, *localisation* within a body system takes precedence over *aetiology*, so when coding a condition which because of its aetiology can be found in several chapters (e.g. trauma), the appropriate chapter should be used.

All non-problem-related care episodes (e.g., family planning, prevention, routine examination) are listed in Chapter A1.

Chapter A (general) should be considered only if the site is not specified or if the disease affects more than two body systems.

Chapters B–W provide specific classes based on the body system or organ involved in the disease and the aetiology.

Conditions accompanying and affecting pregnancy, or the puerperium are usually coded to Chapter W, but a condition is not coded to Chapter W merely because the patient is pregnant; it should be coded to the appropriate class in the chapter representing the body system involved.

All social problems, whether identified as an RFE or as a problem, are listed in Component ZC of Chapter Z.

Problems in activities and participation are listed in Subcomponent 2F0 of Chapter II.

SPECIFIC RULES FOR CODING HEALTH PROBLEMS USING INCLUSION CRITERIA

(See also Chapter 4.)

Rule 1
Coding of diagnoses should occur at the highest level of specificity possible for that patient encounter.

Rule 2
The description contains the information necessary to permit coding to that class.

Rule 3
Consult the description and inclusion after the diagnosis has been formulated. They are **not** guidelines for diagnosis, **nor** are they intended to be used as a guide to therapeutic decisions.

Rule 4
If the description and inclusion do not fit, search in the browser by entering the term(s) in the search box.

Rule 5

For those classes without a description, consult the list of inclusion and index terms in the class and consider any exclusion terms.

DOUBLE CODING

Double coding is only advised for a few classes if recording the manifestation or cause is clinically important. For example, in Class FD67 Retinopathy, it is recommended to also code the known causative agent: such as diabetes TD71 Type 1 diabetes mellitus or TD72 Type 2 diabetes mellitus; or hypertension KD73 Hypertension, uncomplicated or KD74 Hypertension, complicated.

The double coding is advised in the *note* of these classes.

In the ICPC-3 browser, there is an option to search and select more than one code, including extensions of specific codes. These codes can be copied to the clipboard and pasted in a document or electronic system.

CHAPTER 8

Standard for Use of Functioning and Functioning Related

With the Classes from Chapter II, it is possible to describe Functioning and Functioning related aspects of all persons' (first and follow-up) contacts with the health care system in primary and community care settings. The classes acquire meaning when the patient makes a statement about them; for example, if the patient experiences a barrier in participation or a limitation in an activity or if the patient experiences a functioning problem (impairment) in his/her energy, memory or balance.

The classes from Chapter II are person related and do not relate to one EoC specifically. The registration of the Functioning and Functioning related classes can take place both inside and outside the EoC. All registered Functioning classes must always be involved in the analysis of care episodes. Functioning and Functioning related together offer a descriptive 'picture' or 'snapshot' of the person at a certain moment in time. The relation between Functioning and Functioning related and other components can only be understood in the broader context of the ICPC-3 Framework.

SELECTION OF CLASSES

The Functioning and Functioning related items are a selected subset of categories from the WHO International Classification of Functioning, Disability and Health (ICF), which provides an overview of a person in a person-in-context approach, at a certain moment in time.

Where indicated in the references of the classes, a specific set of items is available in the form of a tool for the assessment of functioning (and disability). These sets can be regarded as implementations of the ICF within a specific use case.

- In the first instance there is the World Health Organization Disability Assessment Schedule 2.0 (WHODAS 2.0), which is available at www.psychiatry.org/dsm5. The WHODAS 2.0 is a general tool for the assessment of difficulties due to health/ mental health conditions. This assessment tool is advised for the collection of disability data for adults aged 18 years and older.
- For specific use in primary health care settings, the Primary Care Functioning Scale (PCFS) has been developed and validated for patients of 50 years or older in primary care with chronic morbidity and multi-morbidity. The PCFS is available

in Annex 1. The psychometric properties of the PCFS have been established so it can be used as a valid reliable measurement instrument. Further research with the PCFS is needed to study whether it is also a feasible, efficient and practical instrument for use in the full domain of primary care.[1,2]

- In addition, the Arrêts de Travail en médecine générale à partir de la Classification Internationale de Fonctionnement (ATCIF) has been developed for sick leave prescriptions. In many countries, sick leave prescriptions are frequently used in primary health care/general practice. Using the ICPC-3 for sick leave prescriptions, instead of the traditional medical approach, supports and changes the way health professionals and patients communicate in the work-related context.

The questions from these three questionnaires have been itemised as classes in Chapter II, and their use is encouraged whenever relevant, as separate items or scored with the WHODAS 2.0, the PCFS or the ATCIF.

If greater detail on Functioning and Functioning related aspects is required than that available within the presented selection of items, the WHO ICF should be consulted. Access to the ICF classification is via http://apps.who.int/classifications/icfbrowser/

FUNCTIONING

Functioning of a person can be defined by the complexity of components such as the physiological functions of body systems and psychological functions, anatomical features of parts of the body such as organs, limbs and their components, and the execution of tasks or actions by an individual as such or the involvement of a person in a life situation.

Physiological functions of body systems and psychological functions are referred to as body functions (**body and body system level**).

Anatomical features of parts of the body such as organs, limbs and their components are referred to as body structures (**body level**).

Anatomical features or anatomical structures as such are not classified in the ICPC-3. In the ICPC-3 anatomical terms are harmonised with the Foundational Model of Anatomy, and therefore have the same terminology as in the ICF and the ICD-11.

Execution of tasks or actions by an individual are referred to as Activities (**person level**).

The involvement of a person in a life situation is referred to as Participation (**person-in-social-context level**).

From the primary health care point of view, activities and participation are the core part for shaping a person-centred approach. This means that in the ICPC-3 the Activities and participation subcomponent comes first, followed by the Functions subcomponent.

When the ICPC is used for registering the Functioning component, five rules apply.

Rule 1

The classes from the subcomponent Activities and participation can be coded as an RFE, an EoC/problem, information connected to an RFE and EoC, or as part of linked questionnaires. Without the value score, these classes are of little significance in the context of functioning and should not be used to code Functioning. It is necessary to ask about the degree of severity of the problem if the patient does not express it spontaneously.

Example

'I can't write; I can't hold my pen anymore', 2F25 Fine hand use with the extension PSV.3 complete problem

Rule 2

The classes in the Functions subcomponent from the Functioning component can be used to further explore an RFE or complaint, but may **not** be used as an RFE or EoC. The complaint of dizziness, being tired or being forgetful uttered by the patient is coded with the symptom/complaint classes from the component for organ systems and not with a class from Functioning. Although some class names in Functions overlap with class names in the Symptoms, complaints and abnormal findings component and refer to the same phenomenon, they serve a different purpose or role.

Example

'I am dizzy', RFE NS09

The provider explores dizziness by asking, 'Is it a heavy sensation of rotating or of tilting?'

The patient says, 'it is more rotating but not all the time', 2F83 Dizziness with the extension PSV.1 MILD/MODERATE problem.

'Dizziness' as an impairment (a problem with a function – 2F83) can be used in a descriptive way to understand to what extent a person experiences dizziness as a problem. Without coding the (level of) impairment, the dizziness is just a textual element, which is difficult to trace. Coding as a function makes the dizziness, and the changes in it, traceable, available for discussion and countable.

FUNCTIONING RELATED

Functioning related factors describe the context in which functioning takes place and how functioning is executed. They are made up of Environmental factors (the things outside the person) and Personal factors (how one person differs from another person). Personal factors require the person to express their own perception of their health and the extent to which personal characteristics play a role in the context of their health.

Rule 3
The classes from the Functioning related component are only used to further explore an RFE or complaint or an EoC/problem.

Example

'My daughter has COVID-19; what should I do to avoid becoming infected myself?'
RFE AP203

The patient is asking for advice for the EoC/problem, AP50 Contact with and exposure to communicable diseases.

And if the provider wants to register the living conditions in the context of infection prevention, the code 2R04 Housing should be used with a scale.

Rule 4
Personality functions should only be used if provided by the person her- or himself and with consent for use or reuse. This is not to be used to express the health provider's opinion about the person.

Rule 5
The classes from the Environmental factors subcomponent **are not intended** to code sociodemographic and contextual data. Of course, it can be important to know whether someone lives alone, his/her profession, whether he/she lives in poverty, etc. However, this is more 'background' information; that is, sociodemographic. It describes the context of the patient and has the same value as, for example, age, gender and country of birth.

Example

Not having a paid job does not mean that this is a problem for the patient. If it is a problem for the patient, the class code is ZC17 Unemployment problem. To explore the unemployment problem, the Class 2F58 Remunerative employment can be used with the extension scale. If it is not a problem, the ICPC-3 is not used to register this information.

Case History
A 31-year-old woman comes for an unscheduled visit in the evening. The patient says, 'I have pain in my ankle'. The history is that she had trauma in the

morning. On examination, she has a swollen ankle due to an extensive hematoma. To rule out a possible malleolar fracture, an X-ray is advised. The patient refuses. She is unemployed, belongs to a low social class, is poorly educated and has no health insurance. Furthermore, she is a single mother with a young daughter, aged 12, to look after. In the end, the patient and her doctor agree to put a simple bandage on her ankle, hoping that this will be enough to solve the problem.

Coding this Encounter

The RFE is LS15 Ankle symptom or complaint. During the physical examination, code L102, there are findings that indicate a malleolus fracture. The diagnosis is a possible malleolus fracture, LD36 Fracture of tibia or fibula or both, and the policy is not an X-ray as you would expect, but only a bandage, L211. In this context, registration and coding as not being insured – 2R19 Social security, extension FBV.6 FULL barrier – due to lack of money is an important factor that explains the policy.

In this case, the patient's context is extensively reported: single mother, poorly educated, low social class, unemployed, etc. This information is not related to her refusal to have an X-ray taken. What is directly related is that she is uninsured due to lack of money. And this fact is important for the care offered in this care episode. The other personal context – single mother, poorly educated, low social class, unemployed – is not coded.

FUNCTIONING ASSESSMENT AND THE ICPC

In earlier versions of the ICPC, a class called Limited function/disability (-28) was a standard rubric in every chapter, but almost never used. Functional status measured with COOP/WONCA charts could be coded in this rubric with the addition of an extra digit.[3] But this approach was experienced as problematic, since functional status relates to the patient as a whole and not to the health problem relevant to one chapter specifically. The relationship became difficult to interpret where there was more than one active problem, because comorbidity complicates interpretation.

The ICPC-3 classes in the Functioning and Functioning related components give the health provider the opportunity to describe functioning and functioning-related aspects of all persons' (first and follow-up) contacts. There are references to questionnaires, such as the PCFS, the WHO-DAS 2.0 and the ATCIF, that can be used outside an encounter with a specific RFE or EoC. For instance, the PCFS can be used for all patients over 50 years with multi-morbidity, the WHO-DAS 2.0 can be used as a general tool for the assessment of difficulties due to health/mental health conditions for adults, and the ATCIF can be used for sick leave prescriptions. They are all different applications of ICF classes in the context of the ICPC-3.

Quality of life or overall well-being is not assessed with the ICPC-3. However, as described above, it is possible to describe a person's health-related functioning outside the EoC. In that way, functioning becomes available during every encounter and can inform decision-making, goal setting and outcome measurement.

REFERENCES

1   Postma SAE , van Boven K, ten Napel H, Gerritsen DL, Assendelft WJJ, Schers H, olde Hartman TC. The development of an ICF-based questionnaire for patients with chronic conditions in primary care. *J Clin Epidemiology* 2018; 103: 92–100, Elsevier.

2.   Postma SAE, Schers Henk, Ellis JL, van Boven K, ten Napel H, Stappers Hugo, olde Hartman TC, Gerritsen DL. Primary Care Functioning Scale showed validity and reliability in patients with chronic conditions: a psychometric study, *J Clin Epidemiology* 2020; 125: 130–137.

3.   van Weel C, Konig-Zahn C, Touw-Otten FWMM, van Duijn NP, Meyboom-de Jong B. *Measuring Functional Health Status with the COOP-WONCA Charts: A Manual.* The Hague, CIP-Gegevens Koninklijke Bibiliotheek, 1995.

CHAPTER 9

Standard for Use of Processes of Care (Interventions)

PROCESS OF CARE, INTERVENTIONS

The ICPC can be used to code the interventions used in the process of health care with almost all classes from Chapter I. However, Component -4 Results and some classes of Component -5, Consultation, referral and other reasons for encounter (i.e. -501 Encounter or problem initiated by provider and -502 Encounter or problem initiated by other than patient or provider) cannot be used as an intervention. They can be used as an RFE.

The Process classes are broad and general, rather than specific. For instance, a blood test (-105), even if relating to only one body system (e.g. cardiovascular, K105), may encompass a great variety of different tests, such as of enzymes, lipids or electrolytes.

For components -1, -2 and -6 and the part of Component -5 which can be used to classify the process of care, the class codes are standard throughout the chapters at the three-digit level. The alpha code of the correct chapter must be added by the provider who is doing the coding. Although procedures may not be used as EoCs, there are nevertheless some exceptions, and those are a limited number of rubrics in Chapter W that contain procedures such as delivery and induced abortion.

The following rule for the use of each component of Chapter I reinforces the description of the classes of the components.

> **Rule**
> Whenever a code is shown preceded by a dash (—), select the chapter code from Chapters A–Z. Use Chapter A when no specific chapter can be selected. All codes must begin with an alpha code to be complete. If the episode is a class from Chapter A1, use the component's two-digit alpha code instead of A1.

The most important principle in the coding process is to code all those interventions that take place during the encounter and which have a logical relation to the EoC. For more specificity, a fifth digit may be introduced; see the examples and linkages to the ICHI.

Example 1

—207 Repair/fixation/suture/cast

DOI:10.1201/9781003197157-9

L207.1 Application of casts or the ICHI code PZX.LC.AH
L207.2 Removal of casts

	⟹ **L** stands for the **component Musculoskeletal system**
Convention for	⟹ **207** stands for **Repair/fixation/suture/cast**
coding: L207.1	⟹ **.1** stands for **Application of cast (only)**
ICHI code	⟹ PZX.LC.AH = Application of cast and splint

Example 2

—112 Diagnostic endoscopy
—D112 Diagnostic endoscopy of the digestive system
—D112.1 Gastroscopy or the ICHI code KBF.AE.AD

	⟹ **D** stands for the **Digestive system**
Convention for	⟹ **112** stands for **Diagnostic endoscopy**
coding: D112.1	⟹ **.1** stands for **Gastroscopy (only)**
ICHI code	⟹ KBF.AE.AD = Gastroscopy

More than one Process code may be used for each encounter, but it is extremely important to be consistent. For instance, measuring blood pressure, which is routine for hypertension, can be coded as K102 on every occasion. Routine examinations, complete or partial, both for body systems and for the general chapter must also be coded with consistency. Below are examples of definitions for complete and partial examinations which have been used in one setting. However, it is essential that each country develops a definition of what constitutes a 'complete examination – general' and a 'complete examination – body system' for that culture and that these definitions are used consistently. This will ensure that what is contained in each 'partial examination – general' or 'partial examination – body system', in that country will also have consistency.

COMPLETE EXAMINATION

The term 'complete examination' refers to an examination which contains those elements of professional assessment which, by consensus of a group of local professionals, reflect the usual standard of care. This examination will be complete regarding either the body system (e.g. the eye, Chapter F) or as a complete general examination (Chapter A).

PARTIAL EXAMINATION

The term 'partial examination' in any chapter refers to a partial examination directed to the appropriate specific organ system or function. When more than two systems are involved in a limited or incomplete examination, this is designated general (Chapter A). Most encounters will include a partial examination to evaluate acute and simple illnesses or return visits for chronic illnesses. The following are examples:

Complete examination – general, general check-up -A101

Complete neurological examination -N101
Partial examination – general, limited check on several body systems such as respiratory and cardiovascular and neurological -A102
Partial examination – body system, measuring blood pressure -K102

The following procedures are regarded by the WONCA Classification Committee as being *included in* routine examinations to be coded in rubrics —101 and —102 rather than coded separately:

- inspection, palpation, percussion, auscultation
- visual acuity and fundoscopy
- otoscopy
- vibration sense (tuning fork examination)
- vestibular function (excluding calorimetric tests)
- digital rectal and vaginal examination
- vaginal speculum examination
- blood pressure recording
- indirect laryngoscopy
- height/weight

All other examinations are to be included in other rubrics.

COMPONENT -1 DIAGNOSTIC AND MONITORING INTERVENTIONS

A diagnostic intervention is a clinical intervention intended to diagnose and monitor a patient's disease, condition or injury.

COMPONENT -2 THERAPEUTIC AND PREVENTIVE INTERVENTIONS

Preventive procedures cover a wide range of health care activities, including immunisations, screening, risk appraisal, education, and counselling. Coding of treatment and medications is used to classify those procedures done on site by the primary care provider. It is not intended that this be used to document procedures done by providers to whom the patient has been referred – a much more extensive list of procedures would be required in the latter case.

COMPONENT -3 PROGRAMMES RELATED TO REPORTED CONDITIONS

These care programmes consist of a combination of various interventions, such as asking questions during anamnesis, blood and urine tests, spirometry, advice and policy options, performed in primary care practice.

In general, several health professionals are involved in a 'programme'. This implies that a care plan needs to reflect the integrated approach of all health professionals involved. This could also be referred to as the bio-psycho-social way of working and thinking.

To understand exactly what has been done in the context of the programme, the separate interventions in Component -3 should be coded.

The programmes in Component -3 are already provided with a prefix for the chapters they apply to.

COMPONENT -4 RESULTS

Component -4 does not relate to Process or Interventions.

COMPONENT -5 CONSULTATION, REFERRALS AND OTHER REASONS FOR ENCOUNTER

Consultations and referrals to other primary care providers, physicians, hospitals, clinics or agencies for therapeutic or counselling purposes are to be coded using this component. Also encounters and problems initiated by the provider -501 or by other than the patient or provider -502 are to be coded with classes from this component.

For more specificity, a fifth digit or preferable linkages to locally used referral tables could be added; for example:

- -503 Consultation with a primary care provider
  - -503.1 Nurse
  - -503.2 Physiotherapist
- -505 Referral to other primary care provider
  - -505.1 Nurse
  - -505.2 Physiotherapist
- -506 Referral to specialist, clinic, or hospital
  - -506.1 Internist
  - -506.2 Cardiologist

COMPONENT -6 ADMINISTRATIVE

This component is designed to classify those instances where the provision of a written document or form by the provider for the patient or other agency is warranted by existing regulations, laws or customs. Writing a referral letter is only considered to be an administrative service when it is the sole activity performed during the encounter; otherwise it is included in Component -5. Writing a care plan can be coded here with the rubric code -602.

CHAPTER 10

Standard for Use of Regional Extensions, Emergency Codes and Extension Codes

REGIONAL EXTENSIONS (CHAPTER III)

Although previously the ICPC had been developed to provide a classification for primary health care on an international level, supplementing or completing required data elements in the WHO suite of international classifications, it is also recognised that regional and national primary health care needs must be met. The ICPC-3 has, therefore, extended its content to cater for national and regional coding needs.

In the same manner, the core ICPC-3 codes are based on international frequency, and the Regional extension codes are based on the frequency of classes and codes in national and regional primary care registrations. In addition, classes and codes from the Global Burden of Disease list – needed to achieve a worldwide coverage of health problems – have been included in the Regional extensions. At the moment, there are Regional extensions for Africa, Europe and South America. Extensions for other regions will be available when indicated by the specific region.

In case a request for a new code for a class is submitted by more than two (large) regions, this class, after a thorough update procedure, can be accepted as a four-digit code in the core classification.

In principle, the national or regional classes/codes are part of the inclusions in core classes (chapters A1–II) of the ICPC-3, where the six-digit code is already presented.

Use of the six-digit code is encouraged whenever the specific inclusion term is used. This will prevent the need to invent national codes for terms already in the ICPC-3, and it will support exchange of data.

Examples

'Lassa fever' AD14.05 in the African extension and visible in the core ICPC as inclusion in AD14 Other specified and unknown viral diseases

'scarlet fever' AD24.09 in the European extension and visible in the core ICPC as inclusion in AD24 Other specified and unknown infectious diseases

'Zika virus disease' AD14.08 in the South American extension and visible in the core ICPC as inclusion in AD14 Other specified and unknown viral diseases

DOI:10.1201/9781003197157-10

'hepatitis B carrier' AP80.01 in the African and South-American extension and visible in the core ICPC as inclusion in AP80 Asymptomatic carrier

To prevent the same complaints and illnesses from being assigned different codes in the regional extensions, the application for a new regional code is centrally coordinated.

EMERGENCY CODES (CHAPTER IV)

Chapter IV contains classes with codes for new diseases that can be used in emergency situations of epidemiological importance, especially important given the risk of (national or international) spread of infections. These codes are aligned with ICD codes. In the ICPC-3, there are nine empty classes.

EXTENSION CODES (CHAPTER V)

Extension codes are provided as supplementary codes or additional positions to give more detail or meaning to the initial code if so desired. The extension codes are not to be used without an initial code. In the ICPC-3, there are three categories, of which two apply to specific classes.

SCALE VALUE

Currently, five categories are used:

- The *Consent Scale Value (CSV)* is used by a patient or client to express the level of agreement concerning 2R3 Personality functions. Without these values, the Personality functions (psychic stability, confidence, etc.) have no specific meaning.
- The *Facilitator or Barrier Value (FBV)* is used by a patient or client to express the level of facilitating or acting as a barrier for classes that make up of the environment the person lives in (housing, sanitation, immediate family, etc.).
- The *Forced Expiratory Volume (FEV)* is a calculated ratio for the indication of the volume of air exhaled under forced conditions in the first second of expiration (FEV1). In persons with Chronic Obstructive Lung Disease, it is called the person's vital capacity.
- The *GOLD criteria* or severity scale was developed by the Global Initiative for Chronic Obstructive Lung Disease.
- The *New York Heart Association Functional Classification* is a scale that provides a simple way of classifying the extent of heart failure.
- The *Problem Scale value (PSV)* – in the ICPC-3, no distinction is made between having a problem with a function or a problem with an activity or participation (reading, driving, dressing).

For the Functioning components, the scale values are expressed in terms of the value level of the problem. Using these values at a certain point in time or over a period provides the actual Functioning situation or 'snapshot' of the person. The values can also be used for goal setting and between evaluations of progress.

TEMPORALITY

When indicating the duration of a disorder, a distinction is made between diseases with an acute, subacute and chronic course. The demarcation between the three categories is not clear. Usually, acute conditions last for a period of 4 weeks (WONCA dictionary: less than 4 weeks), subacute conditions for between 1 and 3–6 months, and chronic conditions for longer than 6 months (WONCA dictionary: an illness or disability lasting 6 months or longer).[1]

CAUSALITY

These class types are provided here for informative purposes only to address the causality of classes within a component. A number of these class types have been attributed a specific colour, which is shown in the classification. The colouring is also used for the paper-based desk version to increase the informative value. Causality is indicated in terms of:

- infection
- neoplasm
- trauma
- congenital
- other diagnosis

REFERENCE

1   Bentzen N (ed.). An international glossary for general/family practice. *Fam Pract* 1995; 12: 341–369.

CHAPTER 11

Desk Version, Update Platform and Updates, Licencing of ICPC-3, Translations

DESK VERSION

A desk version with all the classes from the ICPC-3 is available for print. The desk version consists of six pages in A4 format. Just click on the link at the Desk webpage on the ICPC-3.info website. Follow the instructions on how to print the document.

UPDATE PLATFORM AND UPDATES

An update platform is in place for proposed updates. The update platform is part of the ICPC-3 website at the above address. After registration, proposals can be submitted and updates viewed.

After registration and having selected the relevant code, enter the proposal in the lower left window and press 'Add Proposal'. Please make sure to include a reference to the relevant literature supporting the proposal. Alternatively, mail the relevant documentation to info@icpc-3.info.

HOW DOES THE UPDATE PLATFORM WORK?
Instructions

1. Select a four-digit (AA00) class for an update proposal.
2. Specify a specific type of proposal.
3. Explain the suggestion, and indicate the exact code, class name, and ID if available.
4. Specify in detail the motivation for the proposal. Also include a reference to support the update proposal. Without a reference, the update proposal cannot be processed.
5. Consider if the proposal has any consequences for other classes in the ICPC-3, and if this is the case, indicate which code(s) is affected.

After completion, submit the update proposal and, if appropriate, continue with a new update proposal. Update proposals submitted by others can also be viewed.

All fully completed proposal will be reviewed by a team consisting of medical content experts and classification experts and processed further in several steps.

 DOI:10.1201/9781003197157-11

The final date for update proposals is the **first of June** each year. If the update proposal is accepted, it will become effective as of the first of January of the following year.

EMERGENCY UPDATES

The ICPC-3 is a dynamic classification that actively supports primary health care. This means that the ICPC-3 will be updated in a timely manner as needed – in principle, on a yearly basis. If necessary, the ICPC-3 will include classes and codes that are required instantly, such as is the case with pandemics.

TRANSLATIONS

WONCA is an international organisation and wishes to promote versions of ICPC in languages other than English, which is the working language of the ICPC-3 Foundation and the Classification Committee. The ICPC-1 and the ICPC-2 have already been translated into more than 19 languages.

WONCA encourages anyone wishing to promote, assist with, or undertake translations of the ICPC-3 to contact the ICPC-3 Foundation via the website to arrange cooperative work.

A dedicated tool is available for translations, and this provides different output formats, including a thesaurus in the language translated.

The WONCA policy on ICPC-3 translations of the electronic version is as follows:

1. WONCA encourages versions in languages other than English.
2. There must be no changes to the classes. Any extensions must be clearly indicated as such and approved by the WONCA ICPC-3 Foundation prior to publication.
3. Translations must be prepared by named translators working in cooperation with the WONCA ICPC-3 Foundation and to the standards that it sets, particularly in relation to the extent of back translation for checking which may be required.
4. While WONCA will retain the copyright, it will grant without fee the rights to translating organisations to distribute their versions for free. This will require a formal agreement between the WONCA Foundation and the organisation concerned.

POLICY ON COPYRIGHT AND LICENCING

The copyright of the ICPC-3, both in electronic form and hard copy, is owned by WONCA. This policy relates to the electronic version and has the following aims.

Aims

1. to allow WONCA to promote, distribute and support the ICPC-3 and further develop it as the best classification for primary care
2. to maintain international comparability of versions of the ICPC-3
3. to obtain feedback and maintain a clearing house of international experiences with the ICPC-3
4. to achieve recognition of WONCA's initiative and expertise in classification

5. to promote understanding of appropriate links between the ICPC-3 and other classification and coding systems, particularly the ICD-11, the ICD 10, the ICF, and the ICHI
6. to encourage use of the ICPC-3 rather than inhibit it with restrictions
7. to obtain financial support to enable achievement of these aims and allow the work of WONCA to continue and expand

Policy

1. The electronic version of the ICPC-3 should be made available in as many countries as possible. This can be achieved by making it available in the web browser on the ICPC-3 website.
2. Versions involving additions, translations or alterations should be made with input from and agreement of WONCA if they are to be regarded as official WONCA versions are integrated in the Regional extensions of ICPC-3.
3. WONCA should licence appropriate organisations to promote and distribute electronic versions of the ICPC-3 in countries, regions and language groups.
4. Licence fees will be set by negotiation and may be waived when there are advantages to WONCA in so doing, such as when use is for research or development.

Maintenance of the ICPC-3 is made possible by contributions of donors. The members of the earlier ICPC-3 Consortium support future maintenance by annual donations. Licencing the ICPC-3 is a way to expand the services WONCA aims to provide, such as support by translation, implementation, and education.

For more information, consult the ICPC-3.info website.

CHAPTER 12

Tabular List of ICPC-3 Classes

A1 VISITS FOR GENERAL EXAMINATION, ROUTINE EXAMINATION, FAMILY PLANNING, PREVENTION AND OTHER VISITS

Description

The classes in this chapter, like the organ and organ system chapters, are meant to define an episode of care (EoC).

Sometimes there is no apparent health problem involved in an EoC, as, for example, when it relates to need for immunisation or screening, family planning, patient preferences or case finding. These contacts can also be related to first contact or certification.

Interventions carried out in these EoCs are to be coded with the Intervention codes in Chapter I, Interventions and processes; not with the classes in Chapter A1!

AF FAMILY PLANNING
AF01 Procreative management

Description
Encounter for procreative genetic counselling or general procreative counselling, advice on procreation and advice about reversal of previous sterilisation.

Inclusion
female wanting children AF01.00
genetic counselling
male wanting children AF01.01

AF02 Oral contraception

Inclusion
family planning using oral therapy

AF03 Intrauterine contraception

Inclusion
family planning using IUD

AF04 Post-coital contraception

Inclusion
emergency contraception
post-coital intrauterine device AF04.01
morning after pill AF04.00

AF05 Other specified contraception

Inclusion
contraception NOS
contraceptive diaphragm AF05.00
depot contraception AF05.01
sheath contraception, condom AF05.02

AF06 Sterilisation

Inclusion
family planning involving sterilisation

AF99 Other specified family planning

Inclusion
undefined family planning request

AG GENERAL AND ROUTINE EXAMINATION
AG01 General examination and investigation of persons without complaint or reported diagnosis

Inclusion
general medical examination
routine child health examination
routine newborn examination

Exclusion
routine general health check-up of defined subpopulation AG04

AG02 Other specified general examinations and investigations of persons without complaint or reported diagnosis

Inclusion
dental examination
examination of eyes or vision
examination of ears and hearing
examination of blood pressure

AG03 Examination and encounter for certification purposes

Inclusion
determination of paternity
examination for driver license
examination for participation in sports
insurance (life insurance examination)
issue of medical certificate
pre-employment examination

AG04 Routine general health check-up of defined subpopulation

Inclusion
health check-up of armed forces
health check-up of sports teams
occupational health examination
routine check-up for age 60 years and above

AG99 Other specified general and routine examinations

Inclusion
undefined general examination request
undefined routine examination request

AI INTRODUCTION AND PATIENT TREATMENT PREFERENCES

AI01 Introduction to practice and health provider

AI02 Patient treatment and care preferences

Description
Expression or wish to receive or not receive specific treatment or care.

Inclusion
preferences about vaccination
preferences about blood transfusion
preferences about antibiotic treatment
preferences about screening

AI03 Patient preferences about end of life care

Description
Discussion and requests about end of life care encompass more than euthanasia. Other important topics include do-not-resuscitate (DNR) orders, prolonging life with fluids, etc.

AI99 Other specified introduction and patient treatment preferences

Inclusion
blood donor
donor of organs and/or tissue
receiver of blood and/or organs
undefined patient treatment preferences

AP PREVENTION, SCREENING AND CASE FINDING

Description
Episodes of care with a preventive purpose to avoid occurrence or development of a health problem.

AP01 Special screening examination for neoplasms

Inclusion
special screening examination for neoplasm of breast AP01.00
special screening examination for neoplasm of cervix AP01.01
special screening examination for neoplasm of colon and rectum AP01.02
special screening examination for neoplasm of lung AP01.03
special screening examination for neoplasm of prostate AP01.04
special screening examination for neoplasm of skin AP01.05

AP10 Special screening examination for infectious and parasitic diseases

Inclusion
Human Immunodeficiency Virus (HIV) screening
Meticilline Resistant **Staphylococcus aureus** (MRSA) screening
special screening for infections with a predominantly sexual mode of transmission
special screening for intestinal infectious diseases
special screening for tuberculosis

AP15 Special screening examination for diabetes mellitus

AP16 Special screening examination for cardiovascular disorders

AP20 Encounter for immunisation

Exclusion
need for immunisation against influenza AP21
need for immunisation against COVID-19 AP22

AP21 Encounter for immunisation against influenza

AP22 Encounter for immunisation against COVID-19

AP40 Reasons for visit related to lifestyle

Inclusion
assessment of lifestyle
contact with health services for alcohol use
contact with health services for drug use
contact with health services for tobacco use
dietary counselling or surveillance
lifestyle education
lifestyle screening
physical activity assessment

Exclusion
persons encountering health services for other counselling and medical advice AP45

AP45 Persons encountering health services for other counselling and medical advice

Inclusion
counselling related to sexual attitudes
counselling related to sexual lifestyle
counselling related to sexual preference

Exclusion
family planning (persons encountering health services in circumstances related to reproduction) AF

AP50 Contact with and exposure to communicable diseases

Inclusion
contact with and exposure to asymptomatic colonisation by MRSA
contact with and exposure to carrier of infectious disease agent
contact with and exposure to human immunodeficiency virus (HIV)
contact with and exposure to infections with a predominantly sexual mode of transmission
contact with and exposure to tuberculosis

AP60 Potential health hazards related to personal history

Inclusion
immunisation not carried out
personal health surveillance related to personal history
personal history of allergy to drugs, medicaments and biological substances
personal history of malignant neoplasm

personal history of other diseases and conditions
personal history of self-harm
personal history of specific resistance to micro-organisms

Exclusion
polypharmacy care A310

AP65 Potential health hazards related to family history

Inclusion
family history of diabetes AP65.00
family history of ischaemic heart disease AP65.01
family history of malignant neoplasm of breast AP65.02
family history of malignant neoplasm of colon or rectum AP65.03
family history of hypercholesterolaemia AP65.04
family history of malignant neoplasm of other organs
family history of malignant neoplasm of ovary AP65.05
family history of malignant neoplasm of prostate
family history of mental and behavioural disorders
use of di-ethylstilbestrol (DES) by mother AP65.06

AP70 Potential health hazards related to public health

Inclusion
surveillance for infectious diseases
surveillance for any exposure to toxic substances

AP80 Asymptomatic carrier

Description
A carrier is an individual with no overt disease who harbours an infectious organism
or a hereditary chromosome abnormality.

Inclusion
carrier of chromosome disorder AP80.00
carrier, risk for environment or children AP80.02
carrier, risk for him- or herself AP80.03
hepatitis B carrier AP80.01

Exclusion
asymptomatic HIV-infection BD03

AP99 Other specified prevention and case finding

Inclusion
isolation
need for prophylactic surgery

preventive screening and visit
prophylactic immunotherapy
special screening examination for eye and ear disorders
special screening examination for mental and behavioural disorders

AQ PUBLIC HEALTH PROMOTION

Description
Enabling people to increase control over their health and to improve their health. It covers a wide range of social and environmental aspects.

The purpose of health promotion is to positively influence the health behaviour of individuals and communities as well as the living and working conditions that influence their health.

AQ01 Health promotion related to reproductive and sexual health

Description
Guidance and education of individuals and communities related to reproductive and sex-related health behaviour.

AQ02 Health promotion related to growth, development and ageing

Description
Guidance and education of individuals and communities related to growth, development and ageing.

AQ03 Health promotion related to prevention of injury

Description
Guidance and education of individuals and communities related to living and working conditions that influence their health.

AQ04 Health promotion related to prevention of violence

Description
Guidance and education of individuals and communities related to living and working conditions to prevent violence-related health problems.

AQ05 Health promotion related to substance abuse

Description
Guidance and education of individuals and communities related to prevention of substance abuse, narcotic drug abuse and harmful use of alcohol.

AQ99 Other specified health promotion

Inclusion
undefined health promotion request

AR VISITS FOR OTHER REASONS
AR01 Encounter related to presence of devices, implants or grafts

Inclusion
encounter related to presence of pacemaker or implantable cardioverter defibrillator (ICD)

AR02 Encounter related to presence of artificial opening

Inclusion
living with a stoma AR02.00
artificial opening status

AR03 Encounter related to presence of transplanted organ or tissue

Inclusion
status after transplant AR03.00

AR04 Encounter related to postponement of menstruation

Description
Postponement of expected regular menstruation by hormonal treatment.

AR99 Other specified reasons for visit

A GENERAL

AS GENERAL SYMPTOMS, COMPLAINTS AND ABNORMAL FINDINGS
AS01 General pain in multiple sites

Description
Pain is an unpleasant sensory and emotional experience associated with actual or potential tissue damage, or described in terms of such damage. Often, pain serves as a symptom warning of a medical condition or injury. In these cases, treatment of the underlying medical condition is crucial and may resolve the pain. However, pain may persist despite successful management of the condition that initially caused it, or because the underlying medical condition cannot be treated successfully. Chronic pain is pain that persists or recurs for longer than 3 months.

Exclusion
chronic widespread pain (fibromyalgia) LS18

Coding hint
For coding the problem level, consider Pain functions 2F84.

Note
This code should be used only when there is no further specification of site.

AS02 Chills

Description
The sudden sensation of being cold. It may be accompanied by shivering.

Inclusion
rigors
shivers

Exclusion
fever AS03

AS03 Fever

Description
A rise of body temperature above normal.

Inclusion
pyrexia

Exclusion
heat exhaustion/stroke AD45
viral exanthem with fever AD13

AS04 General weakness or tiredness

Description
A sense of decrease in power and energy.

Inclusion
asthenia
exhaustion
fatigue
lassitude
lethargy

Exclusion
drowsiness AS99
heat exhaustion AD45

jetlag AD45
malaise/feeling ill AS06
sleep disturbance PS06

Coding hint
For coding the problem level, consider Energy level 2F71.

AS05 Postviral fatigue

Description
Postviral fatigue is characterised by persistent or recurrent fatigue, diffuse musculoskeletal pain, sleep disturbances and subjective cognitive impairment of 6 months duration or longer. Symptoms are not caused by ongoing exertion; are not relieved by rest; and result in a substantial reduction of previous levels of occupational, educational, social or personal activities. Minor alterations of immune, neuroendocrine and autonomic function may be associated with postviral fatigue.

Chronic fatigue syndrome: considerable cultural variations occur in the presentation of this problem, and two main types occur, with substantial overlap. In one type, the main feature is a complaint of increased fatigue after mental effort, often associated with some decrease in occupational performance or coping efficiency in daily tasks. The mental fatiguability is typically described as an unpleasant intrusion of distracting associations or recollections, difficulty in concentrating and generally inefficient thinking. In the other type, the emphasis is on feelings of bodily or physical weakness and exhaustion after only minimal effort, accompanied by a feeling of muscular aches and pains and inability to relax. In both types a variety of other unpleasant physical feelings is common, such as dizziness, tension headaches and feelings of general instability. Worry about decreasing mental and bodily well-being, irritability, anhedonia and varying minor degrees of both depression and anxiety are all common. Sleep is often disturbed in its initial and middle phases but hypersomnia may also be prominent.

Inclusion
chronic fatigue syndrome AS05.00

Exclusion
weakness/tiredness, general AS04

AS06 Feeling ill

Description
Not in good health.

Inclusion
malaise

Exclusion
cachexia TS07

feeling old PS22
malnutrition TD73

AS07 Fainting

Description
A transient loss of consciousness and postural tone caused by diminished blood flow to the brain.

Inclusion
blackout
collapse
vasovagal attack

Exclusion
coma AS53
feeling faint/giddiness/dizziness NS09

AS09 Swelling and generalised oedema

Inclusion
lump, mass not specified to a location

Exclusion
enlarged lymph gland BS01
oedema KS04
swelling breast (breast lump/mass female) GS26
swelling joint LS20

AS10 Sweating problem

Inclusion
diffuse hyperhydrosis
localised hyperhydrosis AS10.00
night sweats AS10.01
perspiration problem

Exclusion
sweat gland disease SD73

AS11 Bleeding

Exclusion
ecchymosis SD35

Coding hint
Bleeding, haemorrhage just from one site or organ – code to the specific site or organ system.

AS12 Chest pain

Exclusion
pain attributed to chest wall LS04
pain attributed to heart KS01
pain attributed to respiratory system RS01

AS13 Irritable infant

Inclusion
excessively crying infant
restless infant

Exclusion
infantile colic DS01
restless child/adult PS04

AS14 Fall of unknown origin

AS50 Other specified abnormal result investigation

Inclusion
abnormal unexplained hyperglycaemia
abnormal unexplained pathology or imaging results
elevated blood glucose level AS50.00
abnormal thyroid stimulating hormone (TSH) results
subclinical hypothyroidism AS50.01
subclinical hyperthyroidism AS50.02
uraemia

Exclusion
abnormal cervix smear GS50
abnormal urine test US50
raised erythrocyte sedimentation rate BS52
unexplained abnormal white cells BS51
vitamin/nutritional deficiency TD73

AS52 Shock

Description
Shock is a life-threatening medical condition. The circulatory system fails to maintain adequate blood flow, sharply curtailing the delivery of oxygen and nutrients to vital organs.

Inclusion
cardiogenic shock
septic shock
toxic shock syndrome

Exclusion
anaphylactic shock AD46
traumatic shock AD37

AS53 Coma

Description
A prolonged state of deep unconsciousness, often caused by severe injury or illness.

Inclusion
stupor

Exclusion
diabetic coma TD71, TD72
non-diabetic hypoglycaemic coma TD70
syncope AS07

AS90 Concern or fear of disease

Description
Concern about/fear of disease in a patient without the disease, until the diagnosis is proven.

Inclusion
fear of death
fear of dying

Coding hint
If the patient has the disease, code the disease.

AS91 Concern or fear of medical treatment

Inclusion
concern about or fear of the consequences of a drug or medical treatment

Exclusion
adverse effect of drug AD41
complication of medical/surgical treatment AD42

AS92 Concern about appearance

Inclusion
concerns about height
concerns about size
concerns about weight

Exclusion
concern about appearance of breasts GS90
concern about appearance of ears HS91
prominent nose RS91

AS99 Other specified general symptoms, complaints and abnormal findings

Inclusion
clumsiness
cold extremities (acra) AS99.00
drowsiness
drowsy

AD GENERAL DIAGNOSES AND DISEASES
AD01 Measles

Description
Prodrome with infected conjunctivae, fever and cough; plus white specks on a red base in the mucous membranes of the cheek (Koplik's spots), or confluent maculopapular eruption spreading over the face and body, or an atypical exanthem in a partially immune person during an epidemic of measles, or serological evidence of acute measles.

A disease of the respiratory system, caused by an infection with **Morbillivirus**. This disease is characterised by a blotchy rash, fever, cough, conjunctivitis or malaise. This disease may also present with tiny white spots with bluish-white centres inside the mouth. Transmission is by inhalation of infected respiratory secretions, airborne transmission or direct contact. Confirmation is by detection of Morbillivirus RNA or measles-specific IgM antibodies.

Inclusion
complications of measles

Coding hint
generalised rash SS06
viral exanthema AD13

AD02 Chickenpox

Description
A vesicular exanthem which appears in successive crops, with the lesions evolving rapidly from superficial papules to vesicles and eventually to scabs.

A disease caused by an infection with varicella zoster virus. This disease is characterised by a vesicular rash and fever. Transmission is by inhalation of infected respiratory secretions or direct contact with fluid from vesicles.

Inclusion
complications of chickenpox

Exclusion
herpes zoster SD03

AD03 Rubella

Description
An acute exanthem with enlarged lymph nodes, most often suboccipital and post-auricular, with a macular rash on the face, spreading to the trunk and proximal portions of the limbs; or serological evidence of rubella infection.

A disease caused by an infection with the rubella virus. This disease commonly presents with lymphadenopathy or an exanthem that starts on the face and spreads to the limbs and trunk. Transmission is commonly by inhalation of infected respiratory secretions, or direct contact.

Inclusion
complications of rubella

Exclusion
roseola infantum AD13

Coding hint
generalised rash SS06
viral exanthems AD13

AD04 Infectious mononucleosis

Description
Inflammation of the tonsils/pharynx with lymphadenopathy not confined to the anterior cervical nodes, and either atypical lymphocytes on blood smear or splenomegaly; or abnormal heterophile antibody titre or Epstein-Barr virus titre.

A disease typically caused by an infection with Epstein-Barr virus or cytomegalovirus. This disease commonly presents with extreme fatigue, fever, acute pharyngitis, body aches or lymphadenopathy. Transmission is by direct contact with infected body fluids, commonly through saliva.

Inclusion
glandular fever

Coding hint
For coding the problem level, consider Energy level 2F71.

AD13 Other specified and unknown viral exanthems

Inclusion
cowpox
erythema infectiosum (fifth disease) AD13.00
exanthema subitum (sixth disease) AD13.01
fever with rash
hand, foot and mouth disease AD13.02
roseola infantum
unknown viral exanthems

Exclusion
chickenpox AD02
infectious mononucleosis AD04
measles AD01
rubella AD03

AD14 Other specified and unknown viral diseases

Inclusion
adenovirus
chikungunya fever AD14.00
Coxsackie diseases
dengue fever AD14.01
dengue haemorrhagic fever AD14.02
Ebola virus disease AD14.03
hantavirus disease AD14.04
Lassa fever AD14.05
rabies AD14.06
Ross River fever
unknown viral disease
yellow fever AD14.07
Zika virus disease AD14.08

Exclusion
cowpox AD13
erythema infectiosum (fifth disease) AD13
influenza RD07
other viral exanthem AD13

AD15 Tuberculosis

Description
Conversion to a positive tuberculin skin test; or demonstration of **Mycobacterium tuberculosis** on microscopy or culture; or characteristic chest X-ray appearance; or characteristic histological appearance on biopsy.

A disease caused by an infection with the bacteria **Mycobacterium tuberculosis**. This disease presents with symptoms depending on the site of infection. Transmission is commonly by inhalation of infected respiratory secretions.

Inclusion
late effect of tuberculosis
tuberculosis infection of any body site

AD16 Malaria

Description
Intermittent fever with chills and rigors in resident of, or recent visitor to, a malarial region; or demonstration of malarial parasite forms in the peripheral blood.
 A disease caused by an infection with a protozoan parasite from the **Plasmodium** genus. This disease commonly presents with fever, chills, headache, nausea and vomiting, or malaise. Transmission is through the bite of an infected mosquito. Confirmation is commonly by identification of the **Plasmodium** genus in a blood sample.

Inclusion
complications of malaria

AD17 Leishmaniasis

Description
Leishmaniasis is due to infection by vector-borne protozoa from the genus **Leishmania**. Depending on the **Leishmania** species involved, the resultant disease picture may range from a localised cutaneous ulcer through extensive mucocutaneous destruction to severe systemic disease.

Inclusion
cutaneous leishmaniasis AD17.00
mucocutaneous leishmaniasis AD17.01
visceral leishmaniasis AD17.02

AD23 Sepsis

Description
Sepsis as a life-threatening organ dysfunction caused by a dysregulated host response to infection.

Inclusion
urosepsis

Exclusion
puerperal infection or sepsis WD01
sepsis with shock AS52

AD24 Other specified and unknown infectious diseases

Inclusion
African trypanosomiasis AD24.00
brucellosis
Buruli ulcer
Chagas disease (South American trypanosomiasis) AD24.01
filariasis AD24.02
infection caused by **Onchocerca volvulus** AD24.03
infection of unspecified site
leprosy (Hansen's disease) AD24.06
leptospirosis
loiasis (loa loa filariasis) AD24.07
Lyme disease AD24.05
lymphatic filariasis AD24.08
mycoplasma
non-intestinal helminthiases
ornithosis
Q fever
rickettsial disease
scarlet fever AD24.09
toxoplasmosis
unknown infectious disease

Exclusion
meningococcal meningitis ND02
other infection complicating pregnancy/puerperium WD02
perinatal morbidity AD66
puerperal infection/sepsis WD01
viral exanthem, otherwise specified AD13
viral disease, otherwise specified AD14

AD25 Malignancy

Description
Histological evidence of malignancy.

Inclusion
carcinomatosis when primary site is unknown
secondary/metastatic neoplasm when primary site is unknown

Coding hint
disease/condition of unspecified nature/site AD99

AD26 Other specified benign, uncertain or in situ neoplasms

AD35 Multiple trauma and injuries

Inclusion
multiple internal injuries

Note
In this classification 'general' or 'multiple' refers to three or more body sites or systems. Conditions affecting one or two sites should be coded to these sites.

AD36 Other specified and unknown trauma and injury

Inclusion
road traffic accident

Exclusion
fall of unknown origin AS14
late effect of trauma AD37
multiple trauma AD35

AD37 Secondary effect of trauma

Inclusion
deformity or scarring resulting from previous injury
old amputation

Exclusion
post-traumatic stress disorder PD09
psychological effects of trauma/acute stress reaction PS02
scar of skin SD99
wound infection SD07

Coding hint
Code also the nature of the secondary effect of trauma.

AD40 Poisoning by medical agent

Description
Toxicity or impairment produced by accidental or deliberate overdose of an agent which has remedial properties in its usual dosage.

Inclusion
toxic effect overdose of medical agent

Exclusion
insulin coma TD70
medication abuse PS15
suicide attempt PD14

Coding hint
Consider coding the manifestation of the clinical problem (for instance, suicide or suicidal attempt PD13).

AD41 Adverse effect of medical agent

Description
An adverse effect of a medical agent is an undesired harmful effect resulting from a medication. An adverse effect may be termed a 'side-effect' when judged to be secondary to a main or therapeutic effect.

Inclusion
allergy due to medication in proper dose AD41.00
anaphylaxis due to medication in proper dose
drug-induced headache AD41.01
side-effect due to medication in proper dose
spotting using hormonal contraception AD41.02

Exclusion
analgesic nephropathy UD65
contact dermatitis SD70
insulin coma TD70
medication abuse PS15
poisoning by medical agent AD40
reaction to immunisation/transfusion AD42

Coding hint
Symptom or complaint attributed to the proper use of medication, rather than due to disease or injury.

Note
Consider coding the manifestation of the clinical problem.

AD42 Complication of medical treatment

Description
An unexpected and undesired effect resulting from surgical or medical or X-ray treatment or other medical management.

Inclusion
adverse effect of vaccination AD42.00
anaesthetic shock

dehiscence episiotomy AD.42.01
immunisation or transfusion reaction
post-surgical lymphoedema
post-operative infection or haemorrhage or wound disruption
problems due to radiation for diagnosis or treatment

Exclusion
adverse effects of medication AD41
dumping syndrome DD99
hypoglycaemia TD70
poisoning by medical agent AD40
post-gastric surgery syndromes DD99
post-surgical malabsorption, not elsewhere classified DD99

Coding hint
Consider coding the manifestation of the clinical problem.
   In case of Pneumothorax due to surgery, code also RD99.04.

AD43 Side-effect of prosthetic device

Description
Discomfort or impairment or pain or limitation resulting from the fitting or wearing of a device for supplying or amending deficiencies.

Inclusion
side-effect of catheter
side-effect of colostomy
side-effect of gastrostomy
side-effect of heart valve
side-effect of joint replacement
side-effect of organ transplant
side-effect of pacemaker

Exclusion
effect denture/false teeth (prosthetic device) DS19

Coding hint
Consider coding the manifestation of the clinical problem.

AD44 Toxic effect of non-medicinal substance

Description
The nature and effects of chemical (non-medicinal substance), physical or biological poisons on living organisms.

Inclusion
bee sting
general or local toxic effect of carbon monoxide
general or local toxic effect of industrial materials
general or local toxic effect of lead
general or local toxic effect of poisonous animals or insects or plants or snakes
poisoning caused by venomous snake AD44.00
wasp sting

Exclusion
adverse effect medical agent AD41
chronic or acute alcohol abuse (acute alcohol abuse) PS13
chronic or acute alcohol abuse (chronic alcohol abuse) PS12
contact dermatitis SD70
drug abuse PS16
external chemical burns SD41
medication abuse PS15
non-toxic bites/sting insect SD39
non-toxic bites animal/human SD40
poisoning by medical agent AD40
respiratory toxic effects RD99
tobacco abuse PS14

Coding hint
Consider coding the manifestation of the clinical problem.

AD45 Adverse effect of physical factor

Inclusion
adverse effect of cold or lightning or pressure
chilblains AD45.00
drowning
heatstroke and sunstroke AD45.01
hypothermia
jet lag
motion sickness AD45.02

Exclusion
burn due to radiation SD41
effect of alcohol (chronic alcohol abuse) PS12
effect of alcohol (acute alcohol abuse) PS13
effect of medical radiation AD42
effect of tobacco PS14
snow blindness FD36
sunburn SD66

Coding hint
Consider coding the manifestation of the clinical problem.

AD46 Other specified and unknown allergy or allergic reaction

Description
Allergy is a hypersensitivity reaction initiated by a proven immunologic mechanism. Anaphylaxis is a severe, life-threatening systemic hypersensitivity reaction characterised by being rapid in onset with potentially life-threatening airway, breathing or circulatory problems and is usually, although not always, associated with skin and mucosal changes. Also food allergy and angioneurotic oedema.

Inclusion
allergic oedema
anaphylactic shock AD46.00
anaphylaxis
angioneurotic oedema AD46.01
cow's milk protein allergy AD46.02
eggs
food allergy
peanuts
unknown allergy
unknown allergic reaction

Exclusion
allergic rhinitis RD65
allergy resulting from medication AD41
food intolerance DD99
urticaria SD78

AD55 Congenital anomaly, other specified or unknown

Inclusion
chromosome abnormality (Down's syndrome, Marfan's syndrome) and systemic congenital anomalies, not otherwise specified
complete trisomy 21 syndrome AD55.00
congenital rubella
congenital syphilis
unknown congenital anomaly

Coding hint
Anomaly related to a specific body system to be coded to system chapter.

AD65 Premature newborn

Description
Preterm: less than 37 weeks or 259 days gestation.

AD66 Other specified and unknown perinatal morbidity

Description
Morbidity originating in utero or within 7 days of birth.

Inclusion
neonatal sepsis
floppy infant
unknown perinatal morbidity

Exclusion
congenital condition AD55
congenital hydrocephalus ND55
failure to thrive TS08
premature newborn AD65

AD95 Perinatal mortality

Description
Death in utero or within 7 days of birth.

Inclusion
newborn death AD95.00
perinatal and neonatal death
undelivered in utero foetal death AD95.01

AD96 Death

Inclusion
natural death AD96.00
unnatural death AD96.01

Exclusion
perinatal mortality AD95

AD99 Other specified or unknown general diseases or conditions of unspecified site

Inclusion
acquired absence of organs
multi-organ failure

B BLOOD, BLOOD-FORMING ORGANS AND IMMUNE SYSTEM

BS SYMPTOMS, COMPLAINTS AND ABNORMAL FINDINGS OF BLOOD, BLOOD-FORMING ORGANS AND IMMUNE SYSTEM

BS01 Lymph gland(s) symptom or complaint

Description
Enlarged and/or painful lymph nodes.
Enlarged lymph nodes are called lymphadenopathy when there is an abnormal
    enlargement of lymph nodes.

Inclusion
generalised enlarged lymph nodes
localised enlarged lymph nodes
lymphadenopathy with pain
lymphadenopathy without pain

Exclusion
acute lymphadenitis BD01
lymphadenitis, other specified BD02

BS50 Splenomegaly

Description
Splenomegaly is an enlargement of the spleen beyond its normal size.

Exclusion
hypersplenism BD99
hepatomegaly with splenomegaly DS50
splenomegaly with hepatomegaly DS50

BS51 Unexplained changes in white blood cells

Description
Unexplained changes in, or abnormal count of, white blood cells.

Inclusion
persistent or unexplained
neutrophilia unexplained agranulocytosis
unexplained eosinophilia
unexplained leukocytosis
unexplained lymphocytosis
unexplained neutropenia

Exclusion
leukaemia BD25

BS52 Elevated erythrocyte sedimentation rate

Inclusion
red blood cell abnormality

Exclusion
unexplained changes in white blood cell BS51

BS90 Concern or fear of disease of blood, blood-forming organs and immune system

Description
Concern about or fear of other blood or immune system disease in a patient without actually having the disease, until the diagnosis is proven.

Coding hint
If the patient has the disease, code the disease.

BS99 Other specified symptoms, complaints or abnormal findings of blood, blood-forming organs and immune system

Exclusion
splenomegaly BS50

BD DIAGNOSES AND DISEASES OF BLOOD, BLOOD-FORMING ORGANS AND IMMUNE SYSTEM
BD01 Lymphadenitis acute

Description
One or more inflamed or enlarged and tender or painful lymph nodes in the same anatomical location, of recent onset (less than 6 weeks).

Inclusion
abscess of lymph node

Coding hint
enlarged lymph node BS01

BD02 Other specified or unknown lymphadenitis

Description
Enlarged tender lymph nodes present for more than 6 weeks; or demonstration of enlarged inflamed mesenteric lymph nodes by surgery or sonography or lymphography or otherwise.

Inclusion
mesenteric lymphadenitis
unknown lymphadenitis

Exclusion
acute lymphadenitis BD01
acute lymphangitis SD16

Coding hint
enlarged lymph node BS01

BD03 Asymptomatic HIV-infection

Description
Asymptomatic HIV infection confirmed by laboratory criteria according to country definitions and requirements.

BD04 Symptomatic HIV-infection/AIDS

Description
HIV infection and symptomatic clinical stage including severe or stage 4 clinical disease, also known as AIDS, confirmed by laboratory criteria according to country definitions and requirements.

BD25 Malignant neoplasm of blood, blood-forming organs and immune system

Description
Characteristic histological appearance.

Inclusion
Burkitt lymphoma BD25.04
Hodgkin lymphoma BD25.00
leukaemia BD25.02
malignant lymphoma BD25.01
multiple myeloma BD25.03
plasma cell myeloma BD25.03

BD26 Benign, uncertain or in situ neoplasm of blood, blood-forming organs and immune system

Description
Characteristic histological appearance.

Inclusion
benign neoplasm of blood
neoplasm of blood not specified as benign or malignant
polycythaemia rubra vera

Exclusion
malignant neoplasm blood, blood-forming organs and immune system BD25

BD35 Injury of blood, blood-forming organs and immune system

Description
Lesion due to trauma related to blood, blood-forming organs or immune system.

Inclusion
traumatic ruptured spleen BD35.00

BD55 Congenital anomaly of blood, blood-forming organs and immune system

Description
A disease or condition caused by determinants arising in the antenatal period.

Inclusion
congenital anaemia

Exclusion
haemangioma/lymphangioma SD28
haemophilia BD78
hereditary haemolytic anaemia BD65

BD65 Hereditary haemolytic anaemia

Description
A disease caused by a genetically inherited mutation.

Inclusion
haemolytic anaemia due to glucose-6-phosphate dehydrogenase deficiency BD65.00
sickle cell anaemia
sickle cell disorders or other haemoglobinopathies BD65.01
spherocytosis
thalassaemia BD65.02

Exclusion
congenital blood, blood-forming organs and immune system BD55

BD66 Iron deficiency anaemia

Description
Decrease in haemoglobin or haematocrit below levels appropriate for age and sex; plus evidence of blood loss, or microcytic hypochromic red cells by appearance or indices in the absence of thalassaemia, or decreased serum iron and increased iron-binding capacity, or decreased serum ferritin, or reduced haemosiderin in bone marrow, or good response to iron administration.

A disease caused by chronic or acute bleeding, excessive menstrual bleeding, inadequate intake, substances (in diet or drugs) interfering with iron absorption, malabsorption syndromes, inflammation, infection or blood donation. This disease is characterised by decreased levels of iron present in the body. This disease may present with fatigue, pallor or dizziness. Confirmation is by identification of decreased levels of iron in a blood sample.

Inclusion
anaemia due to blood loss

Exclusion
iron deficiency without anaemia TD74

Coding hint
other/unspecified anaemia BD77

BD67 Vitamin B12 anaemia or folate deficiency anaemia

Description
Macrocytic anaemia by smear/indices plus decreased vitamin B12/folate level/positive Schilling test.

Inclusion
folate deficiency anaemia BD67.00
macrocytic anaemia
megaloblastic anaemia due to vitamin B12 deficiency BD67.01
pernicious anaemia

Exclusion
vitamin B12 deficiency without anaemia TD73

BD77 Other specified and unknown anaemias

Inclusion
acquired haemolytic anaemia
aplastic anaemia
blood autoimmune disease
megaloblastic anaemia NOS
protein deficiency anaemia
severe anaemia BD77.00
unknown anaemia

Exclusion
anaemia of pregnancy WD84
iron deficiency anaemia BD66
vitimin B12 anaemia or folate deficiency anaemia BD67

BD78 Coagulation defect

Inclusion
abnormal platelets
haemophilia
hereditary factor VIII deficiency BD78.00
hereditary factor IX deficiency BD78.01
idiopathic thrombocytopenic
immune thrombocytopenic purpura BD78.02
purpura
thrombocytopenia
thrombophilia BD78.03

BD99 Other specified or unknown blood, blood-forming organs, immune system diagnoses or diseases

Inclusion
defects in complement system
hypersplenism
immunodeficiency disorder BD99.00
other haematological abnormality
sarcoidosis BD99.01
secondary polycythaemia

Exclusion
asymptomatic HIV infection BD03
lymphadenitis acute BD01
lymphadenitis chronic/non-specific BD02
lymphoedema KD99
primary inherited erythrocytosis KD99
primary polycythaemia BD26
symptomatic HIV infection BD04

D DIGESTIVE SYSTEM

DS SYMPTOMS, COMPLAINTS AND ABNORMAL FINDINGS OF DIGESTIVE SYSTEM

DS01 General abdominal pain

Inclusion
abdominal colic
abdominal cramps
abdominal discomfort
abdominal pain
acute abdomen
infant colic

Exclusion

biliary colic DD82
dysmenorrhoea GS05
dyspepsia and/or indigestion DS07
epigastric ache DS02
flatulence/gas/belching DS08
heartburn DS03
other localised abdominal pain DS06
renal colic US09

DS02 Epigastric pain

Inclusion

epigastric discomfort
fullness of stomach
stomach ache/pain

Exclusion

dyspepsia and/or indigestion DS07
flatulence/gas/belching DS08

DS03 Heartburn

Description

Substernal pain or burning sensation, usually associated with regurgitation of gastric juice into the oesophagus.

Inclusion

acidity
waterbrash

Exclusion

dyspepsia and/or indigestion DS07
epigastric pain DS02
gastro-oesophageal reflux disease DD67
oesophagitis DD68

DS04 Rectal or anal pain

Inclusion

anal spasm
pain on defaecation
proctalgia fugax

Exclusion

impacted faeces DS12

DS05 Perianal itching

Description
Perianal itching is irritation of the skin at the anal margin and surrounding perianal skin which results in the desire to scratch.

Exclusion
itching SS02
scrotum/testis symptom/complaint GS21

DS06 Other specified localised abdominal pain

Inclusion
colonic pain

Exclusion
abdominal pain, general DS01
biliary colic DD82
dysmenorrhoea GS05
dyspepsia and/or indigestion DS07
epigastric pain DS02
flatulence/gas/belching DS08
heartburn DS03
irritable bowel syndrome DD78
renal colic US09

DS07 Dyspepsia and/or indigestion

Description
A condition characterised by upper abdominal symptoms that suggest indigestion (painful, difficult or disturbed digestion), which may include pain or discomfort of upper abdomen, bloating, feeling of fullness with very little intake of food, nausea and vomiting, heartburn, loss of appetite.

Exclusion
epigastric pain DS02
flatulence/gas/belching DS08
gastro-oesophageal reflux disease DD67
heartburn DS03

DS08 Flatulence, gas and belching

Description
Production or presence of gas in the gastrointestinal tract which may be expelled through the anus and other conditions associated with the production or presence of gas in the GI tract.

Inclusion
bloating
eructation
gas pains
gaseous distension
passing wind

Exclusion
change in abdominal size DS51
dyspepsia/indigestion DS07

DS09 Nausea

Exclusion
alcohol-induced nausea PS13
feelings of overeating DS02
loss of appetite TS03
nausea in pregnancy WS02
vomiting DS10

Note
code for nausea and vomiting as a diagnosis: DS10

DS10 Vomiting

Inclusion
emesis
hyperemesis
retching

Exclusion
haematemesis DS14
vomiting in pregnancy WS02

Note
code for vomiting and diarrhoea as a diagnosis, DS11.

DS11 Diarrhoea

Description
Diarrhoea is an acute or chronic condition in which there is an increased frequency or decreased consistency of bowel movements, usually with excessive and frequent evacuation of watery faeces. Here diarrhoea is described other than specifically described elsewhere such as in motility disorders of intestine or in functional bowel diseases.

Inclusion
frequent or loose bowel movements
watery stools

Exclusion
change in faeces or bowel movements DS18
melaena DS15

DS12 Constipation

Description
Constipation is an acute or chronic condition in which bowel movements occur less often than usual or consist of hard, dry stools that are often painful or difficult to pass. Here constipation is described other than specifically described elsewhere such as in motility disorders of intestine or in functional bowel diseases.

Inclusion
faecal impaction

Exclusion
ileus DD99

DS13 Jaundice

Description
A clinical manifestation of hyperbilirubinaemia of unspecified origin, characterised by the yellowish staining of the skin; mucus membranes and sclera.

Inclusion
icterus
yellow sclera

Exclusion
hematogenous icterus BD77
hemolytic icterus congenital BD65

DS14 Haematemesis

Description
Vomiting of blood that is either fresh bright red or older 'coffee-ground' in character. Vomiting blood is a regurgitation of blood through the upper gastrointestinal tract and it generally indicates bleeding of the upper gastrointestinal tract.

Inclusion
vomiting of blood

Exclusion
haemoptysis RS14

DS15 Melaena

Description
It is bloody stools that indicate bleeding from vascular system in the digestive tract. It is also described as black, tarry and foul-smelling stools or red or maroon-coloured stools that contain degraded blood.

Inclusion
black stools
tarry stools

Exclusion
fresh blood in stool DS16

DS16 Rectal bleeding

Description
Bleeding from anus and anal canal. Bleeding due to specific diseases classified elsewhere (haemorrhoid, cancer, infection, etc.) are excluded here.

Inclusion
anal bleeding
fresh blood in stool

Exclusion
bleeding/haemorrhage AS11
melaena DS15
positive faeces benzidine test AD23

DS17 Incontinence of bowel

Description
Failure of voluntary control of the anal sphincter, with involuntary passage of faeces and flatus.

Inclusion
faecal incontinence

Exclusion
encopresis PS11

DS18 Change in faeces and bowel movements

Description
Bowel habits are the time, size, amount, consistency and frequency of bowel movements throughout the day. A change in bowel habits is any alteration in regular bowel habits. For example, abnormal stool colour, mucous stool or fat in stool.

Exclusion
constipation DS12
diarrhoea DS11
incontinence of bowel DS17
occult blood in stool AD23

DS19 Teeth, gum symptom or complaint

Inclusion
accretions
denture problem
deposits
gingival bleeding
teeth grinding
teething
toothache

Exclusion
caries DD65
teeth and/or gum disease DD65

DS20 Mouth, tongue, lip symptom or complaint

Inclusion
bad breath
coated tongue
cracked lips
dribbling
dry mouth
halitosis
sore mouth
swollen lips

Exclusion
dental/gum problem DS19
cheilosis DD66
disturbance of taste NS08
dehydration TS09

DS21 Swallowing problem

Description
Difficulty in swallowing which may result from neuromuscular disorder or mechanical obstruction. Dysphagia is classified into two distinct types: oropharyngeal dysphagia due to malfunction of the pharynx and upper oesophageal sphincter; and oesophageal dysphagia due to malfunction of the oesophagus.

Inclusion
choking feeling
dysphagia

DS50 Hepatomegaly

Inclusion
hepatomegaly with splenomegaly

DS51 Abdominal distension or abdominal mass or both

Description
This is a condition in which the abdomen feels full and tight because of swelling of the abdomen, usually due to an increased amount of intestinal gas, but occurs sometimes when fluid, substances or mass are accumulating or expanding the abdomen.

Inclusion
abdominal swelling without mass
ascites DS51.00
lump abdomen

Exclusion
flatulence/gas/belching DS08
hepatomegaly DS50
renal mass US09
splenomegaly BS50

DS90 Concern or fear of disease of digestive system

Description
Concern about/fear of disease in a patient without the disease, until the diagnosis is proven.

Coding hint
If the patient has the disease, code the disease.

DS99 Other specified or unknown symptoms, complaints, abnormal findings of digestive system

Inclusion
bruxism

DD DIAGNOSES AND DISEASES OF DIGESTIVE SYSTEM
DD01 Gastrointestinal infection

Inclusion
gastrointestinal infection or dysentery due to amoebiasis DD01.00
gastrointestinal infection or dysentery due to **Campylobacter** DD01.01
gastrointestinal infection or dysentery due to cholera DD01.07
gastrointestinal infection or dysentery due to **Clostridium difficile**
gastrointestinal infection or dysentery due to crytosporidiosis DD01.08
gastrointestinal infection or dysentery due to **Giardia** DD01.02
gastrointestinal infection or dysentery due to norovirus
gastrointestinal infection or dysentery due to rotavirus
gastrointestinal infection or dysentery due to **Salmonella** DD01.03
gastrointestinal infection or dysentery due to **Shigella** DD01.04
gastrointestinal infection or dysentery due to typhoid DD01.05
gastrointestinal infection or dysentery due to **Yersinia enterocolitica** DD01.06

Exclusion
contact with or carrier of infective or parasitic disease AD99
gastroenteritis presumed infection DD05

DD02 Mumps

Description
A disease caused by an infection with mumps virus. This disease commonly presents with fever, headache, fatigue or eventually parotitis. Transmission is by contact with respiratory secretions, directly or indirectly. It is an acute non-suppurative, non-erythematous, diffuse tender inflammation of one or more salivary glands; or acute mumps infection demonstrated by culture or serology; or orchitis in a person exposed to mumps following appropriate incubation period.

Inclusion
mumps meningitis
mumps orchitis
mumps pancreatitis

Coding hint
swelling AS09

DD03 Viral hepatitis

Description
A group of liver diseases caused by infection with one or more of the five hepatitis viruses: hepatitis A virus, hepatitis B virus, hepatitis C virus, hepatitis D virus or hepatitis E virus. Acute infection is defined as recent and present for less than 6 months. Chronic infection is defined as present for more than 6 months, in which

case progression to cirrhosis and liver cancer can occur. Transmission is by the faecal-oral route including water contamination, sexual transmission, blood and body fluid contamination (parenteral spread) and from mother to baby at the time of birth (vertical transmission). Depending on the virus, diagnosis is confirmed by detection of specific viral antigens, anti-viral antibodies or viral nucleic acids in serum.

Inclusion
acute viral hepatitis A DD03.00
acute viral hepatitis B DD03.01
acute viral hepatitis C DD03.02
acute viral hepatitis D
acute viral hepatitis E
chronic viral hepatitis B DD03.03
chronic viral hepatitis C DD03.04
chronic viral hepatitis D DD03.05
chronic viral hepatitis E
all viral hepatitis

Exclusion
carrier of hepatitis virus AP80
other hepatitis DD81

Coding hint
hepatomegaly DS50; jaundice DS13

DD05 Gastroenteritis presumed infection

Inclusion
diarrhoeal disease DD05.00
diarrhoea or vomiting presumed to be infective
dysentery NOS
food poisoning
gastric flu

Exclusion
other specified and unknown diagnoses or diseases of digestive system DD99
irritable bowel syndrome DD78
non-infective enteritis and gastroenteritis (chronic enteritis/ulcerative colitis) DD79

DD06 Perianal abscess

Description
A condition of the anal or rectal region, caused by an infection with a bacterial, viral or fungal source. This condition is characterised by a focal accumulation of purulent material in the anal or rectal region.

Inclusion
ischiorectal abscess

Exclusion
pilonidal abscess SD67

DD07 Intestinal helminths

Description
Either demonstration of helminth in adult form, larvae or ova; or positive skin tests; or positive serology.

Inclusion
ascariasis DD07.00
cyclosporiasis DD07.04
hookworm disease DD07.07
oxyuriasis DD07.01
schistosomiasis DD07.05
strongyloidiasis DD07.06
taeniasis DD07.02

Coding hint
Consider classifying parasitic diseases with the main manifestation outside the digestive system in the other organ chapters.

DD25 Malignant neoplasm of stomach

Description
Characteristic histological appearance.

Inclusion
carcinoma of stomach

Coding hint
other malignant digestive neoplasm (when primary site is uncertain) DD28
benign/unspecified digestive neoplasm DD29

DD26 Malignant neoplasm of large intestine

Description
Characteristic histological appearance.

Inclusion
malignant neoplasm of colon
malignant neoplasm of rectum
malignant neoplasm of anus

Exclusion
familial adenomatous polyposis, DD29 when the histology is not cancer

Coding hint
benign/unspecified digestive neoplasm DD29
other digestive malignant neoplasm (when primary site is uncertain) DD28

DD27 Malignant neoplasm of pancreas

Description
Characteristic histological appearance.

Inclusion
carcinoma of pancreas

Coding hint
benign/unspecified digestive neoplasm DD29
other digestive malignant neoplasm (when primary site is uncertain) DD28

DD28 Other specified or unknown malignant digestive neoplasm

Description
Characteristic histological appearance.

Inclusion
malignant neoplasm of gallbladder/bile ducts DD28.01
malignant neoplasm of lip/mouth/tongue DD28.00
malignant neoplasm of liver DD28.01
malignant neoplasm of oesophagus DD28.02
malignant neoplasm of oral cavity
malignant neoplasm of salivary glands DD28.03
malignant tumour of oropharynx
other specified primary malignancies of digestive system
unknown malignant digestive neoplasm

Exclusion
malignant neoplasm stomach DD25
malignant neoplasm colon/rectum DD26
malignant neoplasm pancreas DD27
secondary malignancy of known site (code to site)
secondary malignancy of unknown site AD25

Coding hint
benign/unspecified digestive neoplasm DD29

DD29 Benign or uncertain neoplasm or carcinoma in situ neoplasm of digestive system

Description
Characteristic histological appearance.

Inclusion
benign digestive neoplasm
digestive neoplasm not specified as benign or malignant when histology is not available
familial polyposis syndrome DD29.00
polyp of colon
polyp of duodenum
polyp of rectum
polyp of stomach

DD35 Injury of digestive system

Inclusion
injury to abdominal organ
injury to teeth
injury to tongue

Exclusion
injury male genital GD35
injury pelvic organs female GD35
laceration skin and/or subcutis SD37
multiple organ injuries AD35

DD36 Foreign body in digestive system

Inclusion
foreign body in digestive tract
foreign body in mouth
foreign body in oesophagus
foreign body in rectum
foreign body swallowed

Exclusion
foreign body in throat/inhaled RD36

DD55 Congenital anomaly of digestive system

Inclusion
biliary anomaly
cleft lip/gum/palate DD55.00
congenital pyloric stenosis DD55.01
Hirschsprung's disease

Meckel's diverticulum DD55.02
Megacolon
oesophageal atresia
tongue-tie DD55.03

Exclusion
congenital metabolic disorder TD56
haemangioma/lymphangioma SD28

DD65 Teeth or gum disease or both

Inclusion
caries
dental abscess
gingivitis DD65.00
malocclusion
temporomandibular joint disorder or syndrome DD65.01

Exclusion
teething and/or denture problem DS19
injury to teeth/gum DD35
Vincent's angina DD66

DD66 Mouth, tongue or lip diseases

Inclusion
angular stomatitis DD66.00
aphthous ulcer
candidiasis of mouth, oral sprue DD66.01
cheilosis
glossitis
mucocele
oral aphthae DD66.02
oral thrush
parotitis
salivary stone DD66.03
stomatitis
Vincent's angina

Exclusion
herpes simplex SD04
mumps DD02
other injury digestive system DD35

DD67 Gastro-oesophageal reflux disease

Description

A condition which develops when the reflux of stomach contents causes troublesome symptoms and/or complications.

Inclusion

gastro-oesophageal reflux disease with oesophagitis DD67.0
gastro-oesophageal reflux disease without oesophagitis DD67.01

Exclusion

oesophagus disease, other specified DD68

DD68 Other specified or unknown oesophagus disease

Inclusion

achalasia
Barrett's oesophagitis DD68.00
benign esophageal stricture DD68.01
Mallory-Weiss syndrome
oesophageal diverticulum DD68.02
oesophagitis
oesophageal ulceration
unknown oesophageal disease
Zenker's diverticulum DD68.03

Exclusion

cancer of oesophagus DD28
gastro-oesophageal reflux disease DD67
hiatus hernia DD74
oesophageal varices KD99

DD69 Duodenal ulcer

Description

Duodenal ulcer is defined as a distinct breach in the mucosa of the duodenum as a result of caustic effects of acid and pepsin in the lumen. Histologically, duodenal ulcer is identified as necrosis of the mucosa extending through the muscularis mucosae into the submucosa. In the endoscopic or radiological view, there is an appreciable depth of the lesion. When the break of epithelial lining is confined to the mucosa without penetrating through the muscularis mucosae, the superficial lesion is called erosion.

Inclusion

bleeding ulcer
duodenal erosion
obstructing ulcer
perforated ulcer

Coding hint
dyspepsia or indigestion DS07
heartburn DS03

DD70 Other specified or unknown peptic ulcer

Description
Characteristic imaging or endoscopy findings, or exacerbation of symptoms in a patient with a previously proven ulcer.

Inclusion
acute erosion
gastric ulcer
gastrojejunal ulcer
ulcus ventriculi DD70.00
unknown peptic ulcer
Zollinger-Ellison syndrome

Exclusion
duodenal ulcer DD69
oesophageal ulcer DD68

Coding hint
dyspepsia/indigestion DS07
heartburn DS03

DD71 Gastritis or duodenitis or both

Description
Gastritis and duodenitis are injuries of mucosa involving epithelial damage, mucosal inflammation and epithelial cell regeneration. This does not include any epithelial defect. Gastritis and duodenitis are caused by various factors such as high acid secretion, infectious agents, drugs, chemical agents or autoimmune reaction. **Helicobacter pylori** can colonise on epithelium and induce gastritis or duodenitis or both.

Inclusion
acute dilatation of stomach
duodenitis
gastritis

Exclusion
gastroenteritis presumed infection DD05
gastrointestinal infection DD01

Coding hint
general abdominal pain DS01
localised abdominal pain DS06
duodenal ulcer DD69

epigastric pain DS02
flatulence/belching DS08
heartburn DS03
indigestion/dyspepsia DS07
nausea DS09
oesophagitis DD67
other specified or unknown peptic ulcer DD70
vomiting DS10

DD72 Appendicitis

Description
Appendicitis is a condition characterised by inflammation of the vermiform appendix.

Inclusion
appendix abscess
appendix perforation

DD73 Inguinal hernia

Description
A hernia occurs when part of an internal organ bulges through a weak area of muscle.
Most hernias occur in the abdomen. Inguinal hernia is the most common type and
is in the groin.

Inclusion
inguinal hernia with incarceration
scrotal hernia

Exclusion
femoral hernia DD76
hydrocele GD71

Coding hint
abdominal mass DS51

DD74 Hiatus hernia

Description
A hernia that occurs through the foramen in the diaphragm.

Inclusion
diaphragmatic hernia

Exclusion
gastro-oesophageal reflux disease DD67
oesophagitis DD68

Coding hint
dyspepsia/indigestion DS07
epigastric pain DS02
heartburn DS03

DD75 Umbilical hernia

Description
A hernia occurs when part of an internal organ bulges through a weak area of muscle. An umbilical hernia is a protrusion of the peritoneum and fluid, omentum or a portion of abdominal organ(s) through the umbilical ring. The umbilical ring is the fibrous and muscle tissue around the navel (bellybutton). Small hernias usually close spontaneously without treatment by age 1 or 2. Umbilical hernias are usually painless and are common in infants.

DD76 Other specified or unknown abdominal hernia

Description
Demonstration of swelling in the specified area and transmitted impulse with cough, or enlargement on straining, or reducible into the abdomen, or intestinal obstruction.

Inclusion
femoral hernia DD76.00
incisional
hernia DD76.01
unknown abdominal hernia
ventral hernia

Exclusion
hiatus hernia DD74
inguinal hernia DD73
umbilical hernia DD75

Coding hint
abdominal mass DS51

DD77 Diverticular disease

Description
Diverticula are a major burden of illness in an ageing population, presenting with bleeding or in form of a diverticulitis. Many are asymptomatic. Most diverticula (pseudodiverticula) occur in the colon; occurrence in the small intestine is also possible, but less frequent.

Inclusion
diverticulitis of intestine
diverticulosis of intestine

Exclusion
Meckel's diverticulum DD55
oesophageal diverticulum DD67

Coding hint
abdominal pain DS01
other localised abdominal pain DS06

DD78 Irritable bowel syndrome

Description
Irritable bowel syndrome (IBS) is a functional bowel disorder in which abdominal pain or discomfort is associated with defaecation or a change in bowel habit and with features of disordered defaecation. The pain can be continuous or an intermittent abdominal pain; and variable bowel pattern over a period of time; and increased gas, or tender and palpable colon, or history of mucous without blood in stool.

Inclusion
spastic colon

Exclusion
allergic/dietetic/toxic gastroenteritis/colitis DD99
gastroenteritis presumed infection DD05
gastrointestinal infection DD01
psychogenic diarrhoea PD10
regional enteritis DD79
vascular insufficiency of gut DD99

Coding hint
abdominal pain DS01
constipation DS12
diarrhoea DS11
flatulence DS08
other localised abdominal pain DS06

DD79 Inflammatory bowel disease

Description
Inflammatory bowel disease is a group of inflammatory conditions of the intestine of unknown aetiology. The pathogenesis is hypothesised that the mucosal immune system shows an aberrant response towards luminal antigens such as dietary factors and commensal microbiota in genetically susceptible individuals.

Inclusion
Crohn's disease (regional enteritis) DD79.00
ulcerative colitis DD79.01

Exclusion
non-ulcerative proctitis DD99

Coding hint
abdominal pain DS01
diarrhoea DS11
mucus colitis DD78

DD80 Anal fissure or anal fistula or both

Description
An anal fissure is a linear break or tear in the mucosa that lines the anal canal. It may occur when hard or large stools are passed after defaecation and typically cause pain and bright red anal bleeding. Anal fistula is an abnormal communication, hollow tract lined with granulation tissue connecting the primary opening inside the anal canal to a secondary opening in the perineal skin. It is usually associated with ano-rectal abscesses, and they are thought to be a chronic condition after an abscess evacuation.

Inclusion
anal fissure DD80.00
fistula ani DD80.01
rectal fistula

Exclusion
perianal abscess DD06

DD81 Other specified or unknown liver diseases

Inclusion
alcohol hepatitis
autoimmune liver disease
cirrhosis of liver DD81.00
fatty liver
hepatitis NOS
liver failure
portal hypertension
steatosis of liver DD81.01
unknown liver diseases

Exclusion
acute viral hepatitis DD03
chronic viral hepatitis DD03
hydatid disease (echinococcosis) DD07

DD82 Cholecystitis or cholelithiasis or both

Description
Inflammation of gallbladder wall by infection of various organism and/or unspecified disorders. Cholelithiasis is calculus of gallbladder, cystic duct or bile duct. Most stones in the gallbladder are asymptomatic, but the most common initial symptom is biliary colic before the development of complications, including acute cholecystitis or cholangitis.

Inclusion
biliary colic
cholangitis DD82.00
cholecystitis DD82.01
cholelithiasis DD82.02
gallstones

Exclusion
primary biliary cholangitis DD81
primary sclerosing cholangitis DD81

DD83 Coeliac disease

Description
Coeliac disease is a permanent intolerance to gluten proteins that are present in wheat, rye and barley. It is an autoimmune disorder, characterised by a chronic inflammatory state of the small intestinal mucosa and submucosa, which can impair digestion and absorption of nutrients, leading to malnutrition.

DD84 Haemorrhoids

Description
Visualisation of varicosities of the venous plexus of the anus or canal, or tender painful blue-coloured localised swelling of acute onset in the perianal area or skin tags in the perianal area.

Inclusion
internal haemorrhoids with or without complications
perianal haematoma
piles
residual haemorrhoidal skin tag
thrombosed external haemorrhoids
varicose veins of anus/rectum

Coding hint
anal lump DS99
anal pain DS04
rectal bleeding DS16

DD99 Other specified or unknown diagnoses or diseases of digestive system

Inclusion
abdominal adhesions
allergic gastroenteropathy
dietetic gastroenteropathy
dumping syndrome
entrapment of intestine in abdominal adhesions DD99.00
food intolerance
ileus DD99.01
intestinal intussusception DD99.02
intestinal obstruction DD99.01
malabsorption syndrome
mesenteric vascular disease
pancreatic disease
pancreatitis DD99.03
peritonitis DD99.04
secondary megacolon
sprue
toxic gastroenteropathy

Exclusion
antibiotic-associated colitis AD41
coeliac disease (non-tropical sprue) DD83
inflammatory bowel disease DD79

F EYE

FS SYMPTOMS, COMPLAINTS AND ABNORMAL FINDINGS OF EYE

FS01 Eye pain

Exclusion
abnormal eye sensations FS07

FS02 Red eye

Inclusion
bloodshot

FS03 Eye discharge

Description
Epiphora is overflow of tears onto the face. A clinical sign or condition that constitutes insufficient tear film drainage from the eyes in that tears will drain down the face rather than through the nasolacrimal system.

Inclusion
epiphora
lacrimation
purulent discharge
watery eye FS03.00

FS04 Visual floaters or spots

Description
Floaters are dark spots or shapes that seem to float in front of the retinal image.

Inclusion
fixed/floating spots in the visual field

FS05 Decreased visual acuity

Description
A decreased vision for sensing form and contour, distant or near, for one or both eyes.

Inclusion
blurred vision
difficulty reading
reduced vision
visual loss
weak eyes

Exclusion
blindness one eye FD72
night blindness FD99
permanent blindness FD72
refractive errors FD69
snow blindness FD36

FS06 Other specified visual disturbances

Inclusion
diplopia
eye strain
photophobia
scotoma and dazzle when symptoms confined to eyes
temporary blindness NOS

Exclusion
night blindness FD99
permanent blindness FD72
refractive errors FD69
snow blindness FD36

FS07 Dry eye or other abnormal eye sensations

Inclusion
burning eye
dry eye (syndrome) FS07.00
itchy eye

Exclusion
eye pain FS01

FS08 Abnormal eye appearance

Inclusion
changed eye colour iris
swollen eye

Exclusion
red eye FS02

FS09 Eyelid symptoms or complaints

Inclusion
abnormal blinking
blepharochalasis FS09.00
ptosis eyelid
xanthelasma palpebrarum FS09.01

Exclusion
inflamed eyelid FD02

FS10 Glasses or contact lenses symptoms or complaints

Inclusion
problems due to spectacles and/or contact lens affecting structure, function or
  sensations of eye(s)

FS90 Concern or ear of eye disease

Description
Concern about/fear of eye disease in a patient without the disease, until the diagnosis
is proven.

Coding hint
If the patient has the disease, code the disease.

FS99 Other specified symptoms, complaints, abnormal findings of eye

Inclusion
abnormal eye movements
nystagmus

FD DIAGNOSES AND DISEASES OF EYE
FD01 Infectious conjunctivitis

Description
Presumed or proven infectious inflammation of conjunctiva.

Inclusion
bacterial conjunctivitis FD01.00
conjunctivitis NOS
viral conjunctivitis FD01.01

Exclusion
allergic conjunctivitis with/without rhinorrhoea FD65
flash burn FD37
other eye inflammation or eye infection FD03
trachoma, chlamydia conjunctivitis FD04

FD02 Blepharitis or stye or chalazion

Description
Generalised and/or localised inflammation and/or swelling of eyelid and/or tarsal gland.

Inclusion
blepharitis FD02.00
chalazion FD02.01
dermatitis of eyelids
dermatosis of eyelids
eyelid infection
hordeolum FD02.02
meibomian cyst
tarsal cyst

Exclusion
dacryocystitis FD03

FD03 Other specified or unknown eye infections or inflammations

Inclusion
dacryocystitis FD03.00
eye infection of unknown cause
eye inflammation of unknown cause
herpes simplex of eye without corneal ulcer
inflammation of orbit
iridocyclitis FD03.01
iritis
keratitis FD03.02

Exclusion
corneal ulcer (herpes) FD05
herpes zoster ophthalmicus SD03
measles keratitis AD01
trachoma FD04

FD04 Trachoma

Description
A disease caused by an infection with the Gram-negative bacteria **Chlamydia trachomatis**. This disease is characterised by a roughening of the inner surfaces of the eyes and inflammation that may lead to superficial vascularisation of the cornea (pannus) and scarring of the conjunctiva. Long-term effects include blindness or other visual impairments. Transmission is by direct or indirect contact with the eyes or nose of an infected individual.

Exclusion
infectious conjunctivitis FD01
other eye infection or inflammation FD03

Coding hint
discharge from eye FS03
red eye FS02

FD05 Corneal ulcer

Description
Loss of epithelial tissue from the surface of the cornea due to progressive erosion and necrosis of the tissue. It is often caused by bacterial, fungal or viral infection.

Inclusion
dendritic ulcer
herpes simplex keratitis dendritic FD05.00
viral keratitis

Exclusion
corneal abrasion/other eye injury FD36

FD25 Neoplasm of eye or adnexa

Inclusion
benign neoplasm of eye/adnexa FD25.00
malignant neoplasm of eye/adnexa FD25.01
uncertain neoplasm of eye/adnexa FD25.02

FD35 Contusion or haemorrhage eye or both

Inclusion
black eye FD35.00
hyphaema
subconjunctival haemorrhage FD35.01

FD36 Other specified and unknown injury of eye

Inclusion
corneal abrasion FD36.00
flash burn
snow blindness FD36.01
unknown injury of eye

Exclusion
contusion or haemorrhage eye FD35
foreign body in eye FD37

FD37 Foreign body in eye

Exclusion
corneal abrasion FD36
congenital stenosis or stricture of lacrimal duct FD55

FD55 Congenital stenosis or stricture of lacrimal duct

Description
This is a condition in which a tear duct has failed to open at the time of birth with an overflow of tears without crying, beginning before the age of 3 months.

Inclusion
congenital dacryostenosis

Exclusion
blocked lacrimal duct in older person FD99
dacryocystitis FD03

FD56 Other specified or unknown congenital anomaly of eye

Inclusion
coloboma
unknown congenital anomaly of eye

Exclusion
congenital stenosis or stricture of lacrimal duct FD55

FD65 Allergic conjunctivitis

Description
Allergic conjunctivitis is an IgE-mediated response due to the exposure of seasonal or perennial allergens in sensitised patients. The allergen-induced inflammatory response of the conjunctiva results in the release of histamine and other mediators. Symptoms consist of redness (mainly due to vasodilation of the peripheral small blood vessels), oedema (swelling) of the conjunctiva, itching and increased lacrimation (production of tears).

Inclusion
acute atopic conjunctivitis
allergic conjunctivitis with rhinorrhoea
allergic conjunctivitis without rhinorrhoea

Exclusion
bacterial/viral conjunctivitis FD01
flash burn FD36
trachoma FD04

FD66 Detached retina

Description
Retinal breaks are full-thickness openings in the neurosensory retina that can be in the form of a hole, a tear or a retinal dialysis. Retinal detachment is a condition in which the retina peels away from its underlying layer of support tissue.

FD67 Retinopathy

Description
Any damage to the retina which may cause visual impairment.

Inclusion
arteriosclerotic retinopathy FD67.00
diabetic retinopathy FD67.01
hypertensive retinopathy

Exclusion
macular degeneration FD68

Note

Double code known causative disease, e.g. diabetes (TD71, TD72) or hypertension (KD73 or KD74).

FD68 Macular degeneration

Description

Degenerative changes in the retina, usually of older adults, which results in a loss of vision in the centre of the visual field (the macula lutea) because of damage to the retina. It occurs in dry and wet forms.

Exclusion

detached retina FD66

FD69 Disorders of refraction and accommodation

Description

Visual deficit correctible with an appropriate lens.

Inclusion

astigmatism FD69.00
hypermetropia FD69.01
long sightedness
myopia FD69.02
presbyopia FD69.03
short sightedness

Exclusion

partial or complete blindness FD72

FD70 Cataract

Description

Cataract is a clouding of the lens inside the eye which leads to a decrease in vision. It is the most common cause of blindness and is conventionally treated with surgery. Visual loss occurs because opacification of the lens obstructs light from passing and being focused on to the retina at the back of the eye.

Inclusion

senile cataract FD70.00

Exclusion

congenital cataract FD56

FD71 Glaucoma

Description
An ocular disease, occurring in many forms, having as its primary characteristics an unstable or a sustained increase in the intraocular pressure which the eye cannot withstand without damage to its structure or impairment of its function.

Inclusion
narrow-angle glaucoma FD71.00
open-angle glaucoma FD71.01
raised intraocular pressure FD71.02
secondary glaucoma FD71.03

Exclusion
congenital glaucoma FD56

FD72 Blindness

Inclusion
partial or complete blindness of both eyes

Exclusion
blurred vision or temporary blindness FS05
colour or night blindness FD99
refractive errors FD69
snow blindness FD36
severe visual impairment

FD73 Strabismus

Description
Lack of parallelism of visual axis of the eyes demonstrated at medical examination.

Inclusion
cross-eye
squint

Coding hint
abnormal eye movement FS99

FD74 Pterygium

Description
Pterygium is a benign growth of the conjunctiva extending onto cornea that is characterised by elastotic degeneration of collagen (actinic elastosis) and fibrovascular proliferation.

FD99 Other specified or unknown diagnosis or diseases of eye and adnexa

Inclusion
amblyopia
arcus senilis
blindness one eye
colour blindness
corneal opacity
disorder of orbit
ectropion FD99.00
entropion FD99.01
episcleritis FD99.02
ingrowing eyelash
lazy eye
night blindness
papilloedema
scleritis FD99.03

G GENITAL SYSTEM

GS SYMPTOMS, COMPLAINTS AND ABNORMAL FINDINGS OF GENITAL SYSTEM

GS01 Pain in penis

Exclusion
priapism or painful erection GS20

GS02 Pain in testis

Inclusion
pain in perineum
pain in scrotum

GS03 Other specified genital pain

Inclusion
pelvic pain
perineal pain
pubic pain
vaginal pain
vulval pain
vulvodynia

Exclusion
breast pain female GS04
dyspareunia female GS23
menstrual pain GS05

GS04 Pain in breast

Inclusion
mastalgia
mastodynia
tenderness of breast

Exclusion
painful breasts in pregnancy or lactation period WS06

GS05 Menstrual pain

Inclusion
dysmenorrhoea
menstrual cramps
menstruation pain

Coding hint
For coding the problem level, consider Pain functions 2F84.

GS06 Intermenstrual pain

Inclusion
mittelschmerz
ovulation pain

GS07 Absent or scanty menstruation

Inclusion
amenorrhoea GS07.00
amenorrhoea primary or secondary
delayed menses
hypomenorrhoea GS07.01
late menses
oligomenorrhoea GS07.02

Exclusion
fear of pregnancy WS90
question of pregnancy WS01

GS08 Excessive menstruation

Inclusion
hypermenorrhoea
menorrhagia
pubertal bleeding

GS09 Irregular or frequent menstruation

Inclusion
frequent menstruation
irregular menstruation
irregular periods GS09.00
metrorrhagia
polymenorrhea GS09.01

Exclusion
menorrhagia or pubertal bleeding GS08

GS10 Intermenstrual bleeding

Inclusion
breakthrough bleeding
dysfunctional uterine bleeding
ovulation bleeding GS10.00
spotting

Exclusion
post-coital bleeding GS15
post-menopausal bleeding GS14

GS11 Premenstrual symptoms or complaints

Description
Symptoms or complaints characterised by cyclic emotional, physical or behavioural symptoms such as mood alterations, psychological changes, fluid retention, neurologic changes, gastrointestinal changes, pelvic heaviness or dermatological changes affecting women in the luteal phase of the menstrual cycle that interfere with an individual's lifestyle.

Exclusion
premenstrual tension syndrome GD68

GS12 Postponement of menstruation

Description
Postponement of expected regular menstruation by hormonal treatment.

GS13 Menopausal symptoms or complaints

Inclusion
atrophic vaginitis GS13.00
menopausal flushing GS13.01
menopausal tension
menopause syndrome
senile vaginitis

Exclusion
postmenopausal bleeding GS14

GS14 Postmenopausal bleeding

Description
A condition of the genital system, caused by polyps, endometrial atrophy, hyperplasia or cancer. This condition is characterised by abnormal uterine bleeding subsequent to the completion of menopause.

GS15 Postcoital bleeding

Inclusion
contact bleeding

GS16 Vaginal discharge

Inclusion
leucorrhoea

Exclusion
atrophic vaginitis GS13
chlamydia genital female GD06
gonorrhoea female GD02
intermenstrual bleeding GS10
urogenital candidiasis female GD06
urogenital trichomoniasis female GD04
vaginal bleeding (menstruation excessive) GS08
vaginal bleeding (menstruation irregular/frequent) GS09

GS17 Other specified vaginal symptoms or complaints

Inclusion
burning in vagina
vaginal dryness
vaginal irritation
vaginal itching
vaginal lesion

vaginal odour
vaginal pruritis

Exclusion
atrophic vaginitis GS13
female genital pain GS03
organic vaginismus GS23

GS18 Vulval symptoms or complaints

Inclusion
labial burning
vulval burning
vulval dryness
vulval itching
vulval irritation

Exclusion
abscess vulva GD69
vulval pain GS03

GS19 Pelvis symptoms or complaints

Exclusion
genital pain female GS03

GS20 Penis symptoms or complaints

Inclusion
foreskin complaint
foreskin symptom
painful erection
priapism

Exclusion
pain in penis GS01
painful ejaculation GS25

GS21 Scrotum or testis symptoms or complaints

Inclusion
lump in testis
swelling of scrotum
swelling of testis GS21.00

Exclusion
pain in testis/scrotum GS02

GS22 Prostate symptoms or complaints

Inclusion
prostatism

Exclusion
urinary frequency and/or urgency US02
urinary retention US04

GS23 Painful intercourse

Inclusion
female dyspareunia
vaginismus

Exclusion
psychogenic sexual problems (sexual desire reduced) PS07
psychogenic sexual problems (sexual fulfilment reduced) PS07

Coding hint
For coding the problem level, consider Sexual functions 2F86.

GS24 Impotence or erectile dysfunction

Description
Male erectile dysfunction is characterised by inability or marked reduction in the ability in men to attain or sustain a penile erection of sufficient duration or rigidity to allow for sexual activity. The pattern of erectile difficulty occurs despite the desire for sexual activity and adequate sexual stimulation, has occurred episodically or persistently over a period of at least several months and is associated with clinically significant distress.

Inclusion
impotence of organic origin
erectile dysfunction

Exclusion
psychogenic impotence or reduced sexual fulfilment PS07
reduced sexual desire PS07

Coding hint
For coding the problem level, consider Sexual functions 2F86.

GS25 Other specified sexual function symptoms or complaints

Inclusion
painful ejaculation

GS26 Lump or mass in breast

Inclusion
lumpy breasts

GS27 Nipple symptoms or complaints

Inclusion
nipple bleeding
nipple cracked
nipple discharge GS27.00
nipple fissure
nipple inversion
nipple pain
nipple pruritus
nipple retraction

Exclusion
nipple symptom or complaint in pregnancy or lactation WS06

GS28 Other specified breast symptoms or complaints

Inclusion
galactorrhoea
gynaecomastia GS28.00
mastopathy

Exclusion
mastitis (lactating) WD03

GS29 Infertility or subfertility

Description
Failure to conceive after 1 year of trying to get pregnant.

Inclusion
primary infertility
secondary sterility

Coding hint
pregnancy symptom or complaint, other WS99

GS50 Abnormal cervix smear

Inclusion
cervical dysplasia
cervical intraepithelial neoplasia (CIN) grade 1
cervical intraepithelial neoplasia (CIN) grade 2

Exclusion
cervical intraepithelial neoplasia (CIN) grade 3 GD32

GS90 Concern about breast appearance

Inclusion
concern about shape of breast
concern about size of breast
dissatisfied with breast appearance

GS91 Concern or fear of sexual dysfunction

Description
Concern about or fear of sexual dysfunction in a patient without sexual dysfunction.

Exclusion
Sexual disfunction PS07

GS92 Concern or fear of sexually transmitted infection

Description
Concern about/fear of sexually transmitted disease in a patient without the disease, until the diagnosis is proven.

Exclusion
fear of HIV/AIDS BS90

Coding hint
If the patient has the disease, code the disease.

GS93 Concern or fear of breast cancer

Description
Concern about/fear of breast cancer in a patient without the disease, until the diagnosis is proven.

Coding hint
If patient has the disease, code the disease.

GS94 Other specified concern or fear of disease of genital system

Description
Concern about/fear of disease in a patient without the disease, until the diagnosis is proven.

Inclusion
fear of prostate cancer GS94.00

Exclusion
concern/fear of breast cancer female GS93
concern/fear of sexual transmitted infection GS92

Coding hint
If the patient has the disease, code the disease.

GS99 Other specified symptoms, complaints and abnormal findings of genital system

Inclusion
haematospermia

Exclusion
urethral discharge US10

GD DIAGNOSES AND DISEASES OF GENITAL SYSTEM
GD01 Syphilis

Description
Demonstration of **Treponema pallidum** on microscopy or positive serological test for syphilis.

Inclusion
condyloma latum
lues
syphilis of any site
urogenital syphilis

GD02 Gonorrhoea

Description
Gonorrhoea is characterised by purulent vaginal, urethral or rectal discharge with Gram-negative intracellular diplococci demonstrated in a patient after a contact with a proven case, or **Neisseria gonorrhoea** cultured.

Inclusion
gonorrhoea of any site

Coding hint
female urethral discharge US10
urethritis UD03

GD03 Genital herpes

Description
Genital herpes is characterised by small vesicles with characteristic appearance and location that evolve into painful ulcers and scabs.

Inclusion
anogenital herpes simplex

GD04 Genital trichomoniasis

Description
Trichomoniasis is a common sexually transmitted infection caused by a parasite. In women, trichomoniasis can cause a foul-smelling vaginal discharge, genital itching and painful urination. Men who have trichomoniasis typically have no symptoms. Pregnant women who have trichomoniasis might be at higher risk of delivering their babies prematurely.

Inclusion
trichomonal vaginitis

Coding hint
vaginal discharge GS16
vaginitis GD12

GD05 Genital human papilloma virus infection

Inclusion
condylomata acuminata
human papilloma virus infection
venereal warts

GD06 Genital Chlamydia infection

Description
An infection with the Gram-negative bacteria **Chlamydia trachomatis**. This infection may be asymptomatic. In females, it may be characterised by fever, painful urination, urinary urgency, dyspareunia, vaginal bleeding or discharge, and pain in the abdomen. In males, it may be characterised by fever, urethritis, painful urination, discharge from the penis, swollen or tender testicles in males. Transmission is by anal, vaginal or oral sex. Confirmation is by identification of **Chlamydia trachomatis**.

Inclusion
cervicitis caused by **Chlamydia** GD06.00
Chlamydia-infection male GD06.03
pelvic inflammatory disease by **Chlamydia** GD06.01
vaginitis caused by **Chlamydia** GD06.02

GD07 Other specified or unknown sexual transmitted disease

Inclusion
lymphogranuloma venerum GD07.00

GD08 Genital candidiasis or balanitis

Inclusion
candida balanitis GD08.00
candidiasis of penis
monilial infection of vagina/cervix
thrush

Exclusion
vaginal discharge GS16
vaginitis GD12

GD09 Pelvic inflammatory disease

Description
Pelvic inflammatory disease is characterised by lower abdominal pain with marked tenderness of uterus or adnexa by palpation, plus other evidence of inflammation.

Inclusion
endometritis
oophoritis
salpingitis

Exclusion
chlamydia infection genital female GD06
genital candidiasis female GD08
genital trichomoniasis female GD04
gonorrhoea female GD02
syphilis female GD01

Coding hint
pelvic congestion syndrome GD69

GD10 Prostatitis or seminal vesiculitis or both

Description
Prostatitis/seminal vesiculitis is characterised by tenderness of prostate/seminal vesicles to palpation and indications of inflammation in urine test.

GD11 Orchitis or epididymitis

Description
Orchitis/epididymitis is characterised by both swelling and tenderness of testes/ epididymis and absence of a specific aetiology (mumps, gonoccocal, tuberculosis, trauma, torsion).

Inclusion
epididymitis GD11.00
orchitis GD11.01

Exclusion
gonococcal orchitis GD02
mumps DD02
torsion of testis GD99
tuberculosis AD15

GD12 Vaginitis or vulvitis

Inclusion
gardnerella
vaginosis (bacterial) GD12.00

Exclusion
atrophic vaginitis GS13
genital candidiasis female GD08
genital trichomoniasis female GD10
trichomoniasis vaginitis GD04

GD25 Malignant neoplasms of cervix

Description
Characteristic histological appearance.

Exclusion
abnormal cervix smear (CIN) grades 1 and 2 GS50
carcinoma-in-situ cervix GD32
cervical intraepithelial neoplasia (CIN) grade 3 GD32

GD26 Malignant neoplasms of prostate

Description
Characteristic histological appearance.

GD27 Malignant neoplasms of breast

Description
Characteristic histological appearance.

Inclusion
adenocarcinoma mammae GD27.00

Exclusion
carcinoma in situ GD32

GD28 Other specified or unknown malignant genital neoplasms

Description
Characteristic histological appearance.

Inclusion
adenocarcinoma of endometrium GD28.00
carcinoma of testis/seminoma
malignant neoplasm of penis GD28.02
malignant neoplasm of testis GD28.03
malignant neoplasm of adnexae
malignant neoplasm of ovaries GD28.01
malignant neoplasm of uterus
malignant neoplasm of vagina
malignant neoplasm of vulva
unknown malignant genital neoplasm

Exclusion
carcinoma in situ GD32

GD29 Fibromyoma of uterus or cervix or both

Description
Fibromyoma of uterus is characterised by enlargement of the uterus not due to pregnancy or malignancy, with single or multiple firm tumours of the uterus/cervix.

Inclusion
fibroid of uterus
fibromyoma of cervix
leiomyoma
myoma of uterus GD29.00

GD30 Benign neoplasms of breast

Description
Characteristic histological appearance.

Inclusion
fibroadenoma of breast

Exclusion
cystic disease of breast GD67

GD31 Benign neoplasms of genital system

Description
Characteristic histological appearance.

Exclusion
benign prostate hypertrophy GD70
physiological cyst of ovary GD69
polyp of cervix GD65

GD32 Genital neoplasm, in situ or uncertain

Inclusion
other carcinoma-in-situ
other genital neoplasm not specified as benign or malignant when histology is not
    available

Exclusion
benign prostatic hypertrophy GD70
endometrial polyp DD29

GD35 Genital injuries

Inclusion
circumcision
corpus alienum genital tract GD35.00
female genital mutilation
foreign body in vagina

Exclusion
genital injury due to childbirth (complicated labour/delivery livebirth) WD82
genital injury due to childbirth (Complicated labour/delivery stillbirth) WD83

GD55 Congenital anomaly of genital system

Inclusion
hermaphroditism
imperforate hymen GD55.00
retractile testis GD55.01

GD56 Hypospadias

GD57 Undescended testicle

Description
This disorder is characterised by the absence of one or both testes from the scrotum.
This disorder may also present with reduced fertility, psychological implications or
increased risk of testicular germ cell tumours.

Inclusion
cryptorchidism
bilateral undescended testicles
unilateral undescended testicle

Exclusion
retractile testis GD55

GD65 Cervical disease

Inclusion
cervical erosion GD65.00
cervical leucoplakia
cervicitis
cervical polyp GD65.01
old laceration of cervix

Exclusion
abnormality of cervix in pregnancy/childbirth/puerperium WD55
abnormal cervix smear GS50

GD66 Uterovaginal prolapse

Description
The descent of one or more of the anterior vaginal wall, posterior vaginal wall, the uterus (cervix) or the apex of the vagina (vaginal vault) or cuff scar after hysterectomy.

Inclusion
cystocele GD66.00
procidentia
rectocele GD66.01

Exclusion
stress incontinence US03

GD67 Fibrocystic disease breast

Description
A condition characterised by changes to the breast tissue leading to benign, non-cancerous lesions in the breast. This condition may be associated with small or large cyst formation, hyperplasia of the ductal epithelium, apocrine metaplasia of the ductal cells, papillomatosis, duct ectasia, sclerosing adenosis or fibrosis of the stroma. This condition may also present with breast pain, thickening of breast tissue, or nipple discharge that worsens prior to menstruation or may be asymptomatic. Confirmation is by clinical breast exam, followed by mammography or ultrasonography to identify abnormal tissue.

Inclusion
chronic cystic disease of breast
cystic fibroadenosis of breast
dysplasia of breast
solitary cyst of breast

GD68 Premenstrual tension syndrome

Description
A syndrome affecting females that is frequently idiopathic. This syndrome is characterised by certain environmental, metabolic, or behavioural factors that occur during the luteal phase of the menstrual cycle, and leads to cyclic emotional, physical, or behavioural symptoms that interfere with an individual's lifestyle.

Coding hint
premenstrual symptom GS11

GD69 Endometriosis

Description
A condition of the uterus that is frequently idiopathic. This condition is characterised by ectopic growth and function of endometrial tissue outside the uterine cavity. This condition may be associated with remaining vestigial tissue from the Wolffian or Mullerian duct, or fragments of endometrium refluxed backward into the peritoneal cavity during menstruation. This condition may also present with dysmenorrhoea, dyspareunia, non-menstrual pelvic pain, infertility, alteration of menses, or may be asymptomatic. Confirmation is by laparoscopy and histological identification of ectopic fragments.

Coding hint
For coding the problem level, consider Pain functions 2F84.

GD70 Benign prostatic hypertrophy

Description
A condition of the prostate, caused by an increased rate of cellular division of the glandular and stromal cells. This condition is characterised by enlargement of the prostatic tissue, dysuria, urinary urgency, nocturia, weak urine stream, straining while urinating, incomplete bladder emptying during urination or increased frequency of urinary tract infection.

Inclusion
hyperplasia of prostate
median bar of prostate
prostatic obstruction
prostatomegaly

Coding hint
dysuria and painful urination US01
incontinence urine US03
other specified urinary problems US05
retention of urine US04
urinary frequency or urgency US02

GD71 Hydrocele or spermatocele or both

Description
A condition characterised by an accumulation of serous fluid (a non-tender fluctuant swelling) in the tunica vaginalis testis or along the spermatic cord, and cystic swelling containing fluid and dead spermatozoa of the testicular epididymis, rete testis or efferent ductuli.

Inclusion
hydrocele GD71.00
spermatocele GD71.01

Coding hint
symptom/complaint of scrotum/testis other GS21

GD72 Phimosis or paraphimosis

Description
Several conditions of the foreskin, caused by abnormalities in the prepuce. This condition is characterised by redundant or tight foreskin and lack of retractability of the foreskin or the inability of the foreskin to be reduced.

GD99 Other specified and unknown diagnoses and diseases of genital system

Inclusion
Bartholin's cyst/abscess GD99.00
epididymal cyst
genital tract fistula female
mastitis (non-lactating) GD99.01
ovarian cyst GD99.02
pelvic congestion syndrome
physiological ovarian cyst
torsion of testis GD99.03

Exclusion
gynaecomastia GS28
mastitis WD03

H EAR

HS SYMPTOMS, COMPLAINTS AND ABNORMAL FINDINGS OF EAR

HS01 Ear pain or ache

Inclusion
otalgia

HS02 Hearing complaint

Inclusion
diplacusis
feeling of hearing loss
hyperacusis
hypoacusis

Exclusion
deafness one ear HD69
deafness both ears HD69
tinnitus HS03

HS03 Tinnitus, ringing or buzzing ear

Description
A non-specific symptom of hearing disorder characterised by the sensation of buzzing,
ringing, clicking, pulsations and other noises in the ear in the absence of appropriate
corresponding external stimuli and in the absence of what the examiner can hear with
a stethoscope.

Inclusion
echo in ear

Exclusion
ears crackling/popping HS99

HS04 Ear discharge

Inclusion
otorrhoea

Exclusion
blood in/from ear HS05

HS05 Bleeding ear

Inclusion
blood from ear
blood in ear

hemorrhage from the ear
otorrhagia

HS06 Plugged feeling in ear

Inclusion
blocked ear

Exclusion
excessive ear wax HD66

HS90 Concern or fear of ear disease

Description
Concern about/fear of ear disease or deafness in a patient without the disease, until
the diagnosis is proven.

Inclusion
fear of deafness

Coding hint
In a patient with the disease, code the disease.

HS91 Concern about appearance of ears

Exclusion
bat ears/congenital anomaly ear HD55

HS99 Other specified symptoms, complaints, abnormal findings of ear

Inclusion
ears crackling
ears popping
itchy ears
pulling at ears

HD DIAGNOSES AND DISEASES OF EAR

HD01 Otitis externa

Description
Inflammation and/or desquamation of the outer ear including the external ear canal.

Inclusion
abscess of external auditory meatus
eczema of external auditory meatus

furuncle of external auditory meatus
abscess and/or eczema and/or furuncle of external auditory meatus

HD02 Acute otitis media or myringitis

Description
Recent perforation of the tympanic membrane discharging pus; or inflamed and bulging tympanic membrane; or one ear drum more red than the other; or red tympanic membrane, with ear pain; or bullae on the tympanic membrane.

Inclusion
acute mastoiditis
acute suppurative otitis media
acute tympanitis
otitis media NOS

Exclusion
chronic otitis media HD05
serous otitis media HD03

Coding hint
ear discharge HS04
ear pain HS01

HD03 Serous otitis media

Description
Visible fluid behind the tympanic membrane, without inflammation; or dullness of the tympanic membrane with either retracting, bulging or with related impairment of hearing.

Inclusion
glue ear
otitis media with effusion (OME)

Exclusion
acute otitis media HD02
chronic otitis media HD05

Coding hint
Eustachian salpingitis/block HD04
plugged feeling ear HS06

HD04 Eustachian salpingitis

Inclusion
Eustachian block
Eustachian catarrh

Eustachian dysfunction
Eustachian tube dysfunction
otosalpingitis
tubotympanitis

Exclusion
serous otitis media HD03

Coding hint
plugged feeling ear HS06

HD05 Chronic otitis media

Inclusion
cholesteatoma HD05.00
chronic mastoiditis
chronic otitis media HD05.01
chronic suppurative otitis media
mastoiditis HD05.02

Exclusion
serous otitis media HD03

HD25 Neoplasm of ear

Inclusion
benign neoplasm of ear HD25.00
malignant neoplasm of ear HD25.01
uncertain neoplasm of ear HD25.02

Exclusion
acoustic neuroma ND25.01
polyp ear HD99

HD35 Acoustic trauma

Description
Noise toxicity can cause hearing loss, either transient or permanent and impairment.
Noise-induced hearing loss typically begins in the high-pitched frequency range of
human voices communication. Deafness in the high-frequency range with a definite
history of exposure to loud noise.

Inclusion
noise deafness

Exclusion
perforation of ear drum HD65
other ear injury HD37

Coding hint
deafness HD69
hearing impairment 2F81

HD36 Foreign body in ear

HD37 Other specified or unknown ear injury

Inclusion
external meatus/pinna injury
traumatic/pressure rupture of ear drum
unknown ear injury

HD55 Congenital anomaly of ear

Inclusion
accessory auricle
bat ears HD55.00
outstanding ears

Exclusion
congenital deafness HD69

HD65 Perforation of ear drum

Inclusion
non-traumatic ruptured ear drum

Exclusion
perforation ear drum with infection (acute otitis media/myringitis) HD02
perforation ear drum with infection (chronic otitis media) HD05
traumatic/pressure rupture ear drum HD37

HD66 Excessive ear wax

Description
Symptom or complaint due to wax in ear canal.

Inclusion
impacted cerumen

HD67 Vestibular syndrome

Description
Syndromes with true rotational vertigo.

Inclusion
benign paroxysmal positional vertigo HD67.00
labyrinthitis HD67.01
Ménière's disease HD67.02
vestibular neuronitis

Coding hint
vertigo/giddiness/dizziness NS09

HD68 Presbycusis

Description
The term presbycusis refers to sensorineural hearing impairment in elderly individuals. Characteristically, presbycusis involves bilateral high-frequency hearing loss associated with difficulty in speech discrimination and central auditory processing of information. Gradual onset with ageing of symmetrical, bilateral deafness, particularly involving high-frequency sounds.

Exclusion
deafness HD69

HD69 Deafness

Inclusion
complete deafness both ears
congenital deafness
deafness one ear
partial deafness both ears

Exclusion
noise deafness HD35
otosclerosis HD99
presbyacusis HD68
temporary deafness H28

HD99 Other specified or unknown diagnoses or diseases of ear and mastoid

Inclusion
constricted external canal
narrow external canal
otosclerosis
polyp of middle ear
stenosis external canal

Exclusion
mastoiditis HD05

K CIRCULATORY SYSTEM

KS SYMPTOMS, COMPLAINTS AND ABNORMAL FINDINGS OF CIRCULATORY SYSTEM

KS01 Pain, pressure, tightness of heart

Inclusion
heaviness of heart
pain attributed to the heart

Exclusion
angina pectoris KD65
chest pain NOS AS12
chest pain attributed to the musculoskeletal system LS04
chest tightness RS99
fear of heart attack KS90
shortness of breath, dyspnoea RS02

KS02 Palpitations, awareness of heart

Inclusion
bradycardia
tachycardia

Exclusion
paroxysmal tachycardia KD69

KS03 Irregular heartbeat

Exclusion
palpitations, awareness of heart KS02

KS04 Ankle oedema

Description
Nearly painless swelling of ankles, usually on both sides, but not always symmetrical.

Inclusion
swollen feet
swollen legs

Exclusion
anasarca/generalised oedema AS09
ankle symptom LS15
localised swelling SS03

KS50 Low blood pressure

Inclusion
idiopathic hypotension

Exclusion
postural hypotension KD75

KS51 Elevated blood pressure

Inclusion
labile hypertension
transient hypertension
white coat hypertension

Exclusion
hypertension, uncomplicated KD73
hypertension, complicated KD74

KS52 Heart murmur or arterial murmur or both

Inclusion
cardiac artery bruit
carotid artery bruit
renal artery bruit
innocent murmur of childhood

Exclusion
cerebrovascular disease ND70
rheumatic heart disease KD02
valve disease KD71

KS90 Concern or fear of disease of circulatory system

Description
Concern about/fear of heart attack or disease in a patient without the disease, until the diagnosis is proven.

Inclusion
fear of heart attack
fear of heart disease
fear of hypertension

Coding hint
If patient has the disease, code the disease.

KS99 Other specified symptoms, complaints, abnormal findings of circulatory system

Inclusion
cardiovascular pain
prominent veins
spider naevus
telangiectasis

Exclusion
cyanosis SS07

KD DIAGNOSES AND DISEASES OF CIRCULATORY SYSTEM

KD01 Infection of circulatory system

Inclusion
acute endocarditis
bacterial endocarditis
chronic endocarditis
myocarditis
pericarditis
subacute endocarditis

Exclusion
arteritis KD99
chronic endocarditis KD71
phlebitis and thrombophlebitis KD78
rheumatic heart disease KD02

KD02 Rheumatic heart disease

Description
For acute rheumatic fever: two major, or one major and two minor manifestations, plus evidence of preceding streptococcal infection; major manifestations: migratory polyarthritis; carditis; chorea; erythema marginatum; subcutaneous nodules of recent onset – minor manifestations: fever; arthralgia; elevated ESR or positive C-reactive protein; prolonged P-R interval on ECG. For chronic rheumatic heart disease: either physical findings consistent with a valve lesion of the heart in a patient with a history of rheumatic fever or physical findings consistent with mitral stenosis, even in the absence of a history of rheumatic fever, but without any other demonstrable cause.

Inclusion
acute rheumatic fever with heart disease KD02.00
acute rheumatic fever without heart disease KD02.01
chorea

Coding hint
heart valve disease KD71
heart disease, other specified KD72

KD25 Neoplasms circulatory system

Inclusion
benign cardiovascular neoplasm KD25.00
malignant cardiovascular neoplasm KD25.01
uncertain cardiovascular neoplasm KD25.02

Exclusion
haemangioma SD28

KD35 Injury of circulatory system

Inclusion
injury of blood vessels

KD55 Congenital anomaly of circulatory system

Inclusion
congenital anomaly of atrial septum KD55.00
Fallot's tetralogy
patent ductus arteriosus
ventricular septal defect KD55.01

Exclusion
haemangioma SD28

KD65 Acute coronary syndrome

Description
Acute coronary syndrome (ACS) is a syndrome, a set of signs and symptoms, due to decreased blood flow in the coronary arteries such that part of the heart muscle is unable to function properly or dies. The most common symptom is chest pain, often radiating to the left shoulder or angle of the jaw, crushing, central and associated with nausea and sweating.

Inclusion
acute myocardial infarction KD65.00
unstable angina pectoris KD65.01

Coding hint
heart pain KS01

Note
double code with KD66

KD66 Chronic ischaemic heart disease

Description
Chronic heart disease is seen due to the atherosclerosis of coronary arteries.

Inclusion
aneurysm of heart
arteriosclerotic heart disease
atherosclerotic heart disease
coronary artery disease
coronary sclerosis KD66.00
ischaemic cardiomyopathy
old myocardial infarction KD66.01
silent myocardial ischaemia
stable angina pectoris KD66.02

Exclusion
acute ischaemic heart disease KD65

KD67 Heart failure

Description
Multiple signs including dependent oedema, raised jugular venous pressure, hepatomegaly in the absence of liver disease, pulmonary congestion, pleural effusion, enlarged heart.

Inclusion
acute heart failure KD67.00
cardiac asthma
chronic heart failure KD67.01
congestive heart failure
diastolic heart failure
left ventricular heart failure
right ventricular heart failure
systolic heart failure

Exclusion
cor pulmonale KD99
pulmonary oedema without heart disease/heart failure RD99

Coding hint
For coding the problem level, consider Energy level 2F71.

KD68 Atrial fibrillation or flutter

Description
Atrial fibrillation is an abnormal cardiac rhythm that is characterised by rapid, unco-ordinated firing of electrical impulses in the upper chambers of the heart. As a result, blood cannot be effectively pumped into the lower chambers of the heart. As in atrial fibrillation, patients with atrial flutter cannot effectively pump blood into the lower chambers of the heart.

Rapid, irregular atrial contractions caused by a block of electrical impulse con-duction in the right atrium and a re-entrant wave front travelling up the inter-atrial septum and down the right atrial free wall or vice versa. Unlike atrial fibrillation, which is caused by abnormal impulse generation, typical atrial flutter is caused by abnormal impulse conduction.

Exclusion
paroxysmal tachycardia KD69

Coding hint
abnormal irregular heartbeat KS03
palpitations KS02
paroxysmal atrial fibrillation

KD69 Paroxysmal tachycardia

Description
History of recurrent episodes of rapid heart rate (over 140 beats per minute) with both abrupt onset and termination.

Inclusion
re-entry tachycardia
supraventricular tachycardia KD69.00
ventricular tachycardia KD69.01

Exclusion
tachycardia NOS KS02
atrial fibrillation KD68

Coding hint
abnormal irregular heartbeat KS03
palpitations KS02

KD70 Cardiac arrhythmia or conduction disorder or both

Description
One or more heartbeats which occur at times other than the regular beats of the underlying rhythm.

Inclusion
atrial premature beats
atrioventricular block KD70.00
bigeminy
bundle branch block
cardiac arrhythmia
conduction disorder
ectopic beats
extrasystoles
heart block
junctional premature beats
left bundle-branch block
long Q-T syndrome KD70.01
other conduction disorders
premature beats
right bundle branch block
sick-sinus syndrome KD70.02
supraventricular extrasystoles KD70.03
ventricular extrasystoles KD70.04
ventricular fibrillation/flutter
ventricular premature beats
Wolff-Parkinson-White syndrome KD70.05

Exclusion
paroxysmal tachycardia KD69

Coding hint
abnormal irregular heartbeat KS03
palpitations KS02

KD71 Heart valve disease

Description
Evidence of valvular dysfunction by either characteristic heart murmur or by imaging/ echocardiographic evidence of abnormal valve.

Inclusion
cardiac valve prolapse KD71.00
mitral valve insufficiency or incompetence or mitral regurgitation KD71.01
mitral valve prolapse
non-rheumatic aortic/mitral/pulmonary/tricuspid valve disorder
stenosed aortic valve KD71.02

Exclusion
rheumatic valve disease KD02

Coding hint
cardiac murmur NOS KS52
hypertensive heart disease KD74

KD72 Other specified and unknown heart disease

Inclusion
cardiac arrest KD72.00
cardiac arrest with successful resuscitation KD72.01
cardiomegaly
cardiomyopathy KD72.02
non-infectious disease of pericardium
non-infectious myocarditis
unknown heart disease

KD73 Hypertension, uncomplicated

Description
Although a continuous association exists between higher blood pressure and increased cardiovascular disease risk, it is useful to categorise blood pressure levels for clinical and public health decision-making. Look at the guidelines for the criteria. The complications of uncontrolled or prolonged hypertension include damage to the blood vessels, heart, kidneys and brain.

Inclusion
essential hypertension
idiopathic hypertension

Exclusion
hypertension with complications KD74
hypertension in pregnancy WD70

Coding hint
elevated blood pressure KS51

Note
For children, consult appropriate paediatric blood pressure tables.

KD74 Hypertension, complicated

Description
The complications of uncontrolled or prolonged hypertension include damage to the blood vessels, heart, kidneys and brain.

Inclusion
hypertensive heart disease
hypertensive renal disease

malignant hypertension
secondary hypertension

Exclusion
uncomplicated hypertension KD73

Note
1. For children, consult appropriate paediatric blood pressure tables.
2. If secondary hypertension, code also the underlying cause.

KD75 Postural hypotension

Description
Signs or symptoms of cerebrovascular insufficiency (dizziness, syncope) on changing
from the supine to the upright position; and a fall in mean blood pressure of 15 mmHg
on two or more occasions when changing from the supine to the upright position.

Inclusion
orthostatic hypotension

Exclusion
hypotension due to drugs AD41
low blood pressure KS50
ideopathic hypotension KS50

Coding hint
low blood pressure KS99

KD76 Atherosclerosis or peripheral vascular disease

Inclusion
arterial embolism
arterial thrombosis
arterial stenosis
arteriosclerosis
atherosclerosis
atheroma
Buerger's disease
endarteritis
gangrene
intermittent claudication KD76.00
limb ischaemia
Raynaud's syndrome KD76.01
thromboangiitis obliterans KD76.02
vasospasm

Exclusion
acute myocardial infarction KD65
aneurysm KD99
cerebral atherosclerosis (transient cerebral ischaemia) ND68
cerebral atherosclerosis ND70
ischaemic heart disease with angina KD65
ischaemic heart disease without angina KD66
mesenteric atherosclerosis DD99
ophthalmic/retinal atherosclerosis FD99
pulmonary atherosclerosis KD77
stroke/cerebrovascular accident ND69
renal atherosclerosis UD99

KD77 Pulmonary embolism

Description
This is a blockage of the main artery of the lung or one of its branches by a substance that has travelled from elsewhere in the body through the bloodstream (embolism) with a sudden onset of dyspnoea/tachypnoea and either clinical or imaging evidence of pulmonary infarction or ECG evidence of acute right ventricular strain.

Inclusion
pulmonary (artery/vein) infarction
pulmonary thromboembolism
pulmonary thrombosis

Coding hint
chest pain AS12
dyspnoea RS02

KD78 Thrombosis or phlebitis or thrombophlebitis

Inclusion
deep vein thrombosis KD78.00
phlebothrombosis
portal thrombosis
superficial vein thrombophlebitis KD78.01
superficial vein thrombosis

Exclusion
cerebral thrombosis ND69, ND70

KD79 Varicose veins

Description
Presence of dilated superficial veins or demonstration of valve incompetence of veins.

Inclusion
scrotal varices/varicocele KD79.00
varicose eczema
varicose veins of sites other than lower extremities
varicocele
venous insufficiency KD79.01
venous stasis

Exclusion
oesophageal varices KD99
varicose ulcer SD77
varicose veins of anus/rectum DD84

Coding hint
prominent veins KS99

KD99 Other specified and unknown diagnoses and diseases of the circulatory system

Inclusion
aortic aneurysm or dissection KD99.00
arteriovenous fistula
arteritis
arteritis temporalis KD99.03
diabetic peripheral angiopathy KD99.01
lymphoedema KD99.04
oesophageal varices KD99.02
other aneurysm polyarteritis nodosa
vasculitis

Exclusion
atherosclerotic arterial stricture/stenosis KD76
cerebral aneurysm ND70
chronic/non-specific lymphadenitis BD02
gangrene KD76

L MUSCULOSKELETAL SYSTEM

LS SYMPTOMS, COMPLAINTS AND ABNORMAL FINDINGS OF MUSCULOSKELETAL SYSTEM
LS01 Neck symptom or complaint

Inclusion
cervicalgia
neck symptom

pain attributed to cervical spine
neck stiffness

Coding hint
Consider the syndrome LD65.
For coding the problem level, consider Pain functions 2F84.
Use the extension codes for distinguishing between acute and chronic.

LS02 Back symptom or complaint

Inclusion
backache
dorsalgia
thoracic back pain

Exclusion
low back pain LS03

Coding hint
Consider the syndrome LD66 or LD67.
For coding the problem level, consider Pain functions 2F84.
Use the extension codes for distinguishing between acute and chronic.

LS03 Low back symptom or complaint

Inclusion
coccydynia
lumbago
lumbalgia
lumbar and sacroiliac back pain

Exclusion
sciatica LD67
thoracic back pain LS02

Coding hint
Consider the syndrome LD66 or LD67.
For coding the problem level, consider Pain functions 2F84.
Use the extension codes for distinguishing between acute and chronic.

LS04 Musculoskeletal chest symptom or complaint

Inclusion
chest pain attributed to musculoskeletal system
intercostal pain
swelling on chest

Exclusion
chest pain AS12
intercostal neuralgia ND77
pain attributed to the heart KS01
painful respiration/pleuritic pain/pleurodynia RS01

LS05 Flank or axilla symptom or complaint

Inclusion
flank pain
loin pain
pain in axilla

Exclusion
kidney symptom US07

LS06 Jaw symptom or complaint

Inclusion
temporomandibular joint symptom

Exclusion
teeth/gum symptom/complaint DS19
temporomandibular joint disorder DD65

Coding hint
Consider the syndrome DD65.

LS07 Shoulder symptom or complaint

Coding hint
Consider the syndrome LD68.

LS08 Arm symptom or complaint

Exclusion
muscle pain/myalgia LS17

LS09 Elbow symptom or complaint

LS10 Wrist symptom or complaint

LS11 Hand or finger (or both) symptom or complaint

Inclusion
cramp in hands
pain in joint of hand or finger
pain in fingers
pain in hand

LS12 Hip symptom or complaint

LS13 Leg or thigh (or both) symptom or complaint

Inclusion
buttock pain
leg cramps
leg weakness

Exclusion
growing pain LS99
muscle pain/myalgia LS17
restless legs NS03

LS14 Knee symptom or complaint

Inclusion
effusion or swollen knee

LS15 Ankle symptom or complaint

Exclusion
ankle oedema KS04

LS16 Foot or toe (or both) symptom or complaint

Inclusion
foot/feet cramp
heel pain
metatarsalgia LS16.00

LS17 Muscle pain

Inclusion
abdominal wall pain
myalgia
rheumatism

Exclusion
pain in neck LS01
pain in back LS02
pain in lower back LS03
leg cramps LS13

LS18 Chronic widespread pain

Description
Chronic widespread pain (CWP) is diffuse pain in at least 4 of 5 body regions and is
associated with significant emotional distress (anxiety, anger/frustration or depressed

mood) and functional disability (interference in daily life activities and reduced participation in social roles). CWP is multifactorial: biological, psychological and social factors contribute to the pain syndrome. The diagnosis is appropriate when the pain is not directly attributable to a nociceptive process in those regions and there are features consistent with nociplastic pain and identified psychological and social contributors. Other chronic pain diagnoses to be considered are chronic cancer pain, chronic post-surgical or post-traumatic pain, chronic neuropathic pain, chronic visceral pain and chronic musculoskeletal pain.

Inclusion
fibromyalgia
fibromyositis
primary fibromyalgia syndrome LS18.00

Coding hint
For coding the problem level, consider Pain functions 2F84.

LS19 Muscle symptom or complaint

Inclusion
atrophy of muscle
muscle stiffness
muscle strain
wasting of muscle
weakness of muscle

Exclusion
'growing pains' in child LS99
leg cramps LS13
muscle pain LS17
pain in neck LS01
pain in back LS02
pain in lower back LS03
restless legs NS03

LS20 Other specified joint symptoms or complaints

Inclusion
arthralgia
effusion of other specified joint
multiple joint symptoms or complaints
pain in joint
stiffness in joint
swelling of joint
weakness in joint

Exclusion
ankle symptom/complaint LS15
elbow symptom/complaint LS09
foot/toe symptom/complaint LS16
hand/finger symptom/complaint LS11
hip symptom/complaint LS12
jaw symptom/complaint LS06
knee symptom/complaint LS14
shoulder symptom/complaint LS07
wrist symptom/complaint LS10

LS90 Concern or fear of disease of musculoskeletal system

Description
Concern about/fear of disease of musculoskeletal system in a patient without the disease, until the diagnosis is proven.

Coding hint
If the patient has the disease, code the disease.

LS99 Other specified symptoms, complaints and abnormal findings of musculoskeletal system

Inclusion
abnormal posture
'growing pains' in a child

Exclusion
clubbing of fingernails SS09

LD DIAGNOSES AND DISEASES OF MUSCULOSKELETAL SYSTEM
LD01 Infection of musculoskeletal system

Description
Infection localised in musculoskeletal system.

Inclusion
bacterial (septic) arthritis LD01.00
infective tenosynovitis
osteomyelitis LD01.01
pyogenic arthritis

Exclusion
Reiter's disease LD99
late effect of polio ND01

LD25 Malignant neoplasm musculoskeletal system

Description
Characteristic histological appearance.

Inclusion
fibrosarcoma
osteosarcoma

Exclusion
benign/unspecified musculoskeletal neoplasm LD26
secondary neoplasms (code to original site)

LD26 Benign, uncertain or carcinoma in situ musculoskeletal

Inclusion
benign musculoskeletal neoplasm
muscoloskeletal neoplasm in situ
musculoskeletal neoplasm not specified as benign or malignant when histology is not
    available
osteochondroma

Exclusion
malignant musculoskeletal neoplasm LD25

LD35 Fracture of radius or ulna or both

Description
Imaging evidence of a fracture; or trauma plus visible/palpable deformity or crepitus
involving the bone.

Inclusion
Colles' fracture
elbow fracture

Exclusion
non-union LD99

Coding hint
If it is a pathological fracture, code also the underlying disease.

LD36 Fracture of tibia or fibula or both

Description
Imaging evidence of a fracture; or trauma plus visible/palpable deformity or crepitus
involving the bone.

Inclusion
Pott's fracture

Exclusion
fracture patella LD39
non-union LD99

Coding hint
If it is a pathological fracture, code also the underlying disease.

LD37 Fracture of hand or foot bone or both

Description
Imaging evidence of a fracture; or trauma plus visible/palpable deformity or crepitus involving the bone.

Inclusion
fracture of carpal bone
fracture of metacarpal bone
fracture of phalanx hand LD37.00
fracture of phalanx foot LD37.01
fracture of tarsal bone
fracture of metatarsal bone

Exclusion
non-union LD99

Coding hint
If it is a pathological fracture, code also the underlying disease.

LD38 Fracture of femur

Description
Imaging evidence of a fracture; or trauma plus visible/palpable deformity or crepitus involving the bone.

Inclusion
fracture of neck of femur LD38.00

Exclusion
non-union LD99

Coding hint
If it is a pathological fracture, code also the underlying disease.

LD39 Other specified and unknown fracture

Description
Imaging evidence of a fracture; or trauma plus visible/palpable displacement of the bone surface.

Inclusion
fracture of clavicle LD39.01
fracture of humerus LD39.02
fracture of nasal bones LD39.00
fracture of pelvis LD39.05
fracture of patella LD39.06
fracture of rib LD39.03
fracture of skull LD39.07
fracture of vertebral column LD39.04
unknown fracture

Exclusion
fractures in radius/ulna LD35
fractures in tibia/fibula LD36
fractures in hand/foot bone LD37
fractures in femur LD38
fractured skull with cerebral injury ND36
non-union LD99

Coding hint
If it is a pathological fracture, code also the underlying disease.

LD45 Injury to multiple structures of knee

Description
An initial injury which occurred no longer than 1 month previously and demonstration of ligament/meniscus tear by surgery/arthroscopy/imaging, or by locking/giving way, pain and swelling of knee. Or a stretch injury of the affected part plus pain aggravated by stretching or tensing the affected structure.

Inclusion
acute damage to meniscus/cruciate ligaments
acute damage to collateral ligaments of knee
acute (traumatic) derangement of knee LD45.00
rupture of cruciate ligaments LD45.01
sprain of cruciate ligaments of knee LD45.01
sprain of lateral collateral ligament of knee LD45.02
sprain of medial collateral ligament of knee LD45.02
tear of meniscus of knee LD45.03

Exclusion
chronic internal damage to knee LD99
dislocation of patella LD48

LD46 Sprain or strain of ankle

Description
A stretch injury of the affected part plus pain aggravated by stretching or tensing the affected structure.

LD47 Other specified and unknown sprain or strain of joint

Description
A stretch injury of the affected part plus pain aggravated by stretching or tensing the affected structure.

Inclusion
sprain/strain of other joint/ligament
unknown sprain or strain of joint
whiplash injury of neck LD47.00

Exclusion
sprain/strain ankle LD46
sprain/strain knee LD45
back strain LD66
cervical neck sprain LD65

LD48 Dislocation or subluxation

Description
A trauma to the joint plus either imaging evidence of a dislocation/subluxation or visible/palpable dislocation deformity.

Inclusion
closed subluxation of jaw LD48.00
dislocation acromioclavicular of joint LD48.01
dislocation of any site, including spine
dislocation of finger LD48.02
dislocation of shoulder joint LD48.03
open dislocation of jaw LD48.00
subluxation acromioclavicular of joint LD48.01
subluxation of any site, including spine
subluxation of finger LD48.02
subluxation of radial head LD48.04
subluxation of shoulder joint LD48.03

Coding hint
Code fracture dislocations to the fracture.

LD49 Other specified musculoskeletal injury

Inclusion
contusion of rib LD49.00
deep foreign body
tear musculus gastrocnemius LD49.01
traumatic amputation
traumatic haemarthrosis

Exclusion
animal bite SD40
bruise/contusion SD35
head injury/concussion/intracranial injury/skull fracture ND36
injury teeth DD35
injury eardrum HD65
insect bite/sting SD39
internal injury of chest/abdomen/pelvis, multiple trauma AD35
laceration/open wound SD37
laceration/other injury to nerve ND37
late effect trauma/deformity/disability/scarring AD37
non-/mal-union of fracture LD99
traumatic arthropathy LD80

LD55 Congenital anomaly of musculoskeletal system

Inclusion
bow leg
cervical rib LD55.00
clubfoot (talipes)
congenital dislocation of hip LD55.01
congenital hip dysplasia LD55.01
congenital malformation of skull and face
genu recurvatum
other congenital deformity of the foot
spina bifida occulta LD55.02
talipes equinovarus LD55.03

Exclusion
pes planus (acquired) LD71
scoliosis LD70
spina bifida ND55

LD65 Neck syndrome

Description
Cervical pain from the neck, with or without radiation.

Inclusion
cervical disc lesion with/without radiation of pain
cervical herniation of nucleus pulposus LD65.00
cervicobrachial syndrome with/without radiation of pain
cervicogenic headache with/without radiation of pain
osteoarthritis of neck with/without radiation of pain
radicular syndrome of upper limbs with/without radiation of pain
spondylosis with/without radiation of pain
torticollis with/without radiation of pain

Exclusion
whiplash injury of neck LD47

Coding hint
For coding the problem level, consider Pain functions 2F84.

LD66 Back syndrome without radiating pain

Description
Back pain without radiation plus limitation of movement confirmed at medical examination.

Inclusion
back strain
collapsed vertebra
facet joint degeneration
osteoarthrosis or osteoarthritis of spine
spondylolisthesis LD66.01
spondylosis LD66.00
spondylolysis LD66.01

Exclusion
back pain with radiation/sciatica LD67
coccydynia LS03
syndrome related to the neck LD65

Coding hint
For coding the problem level, consider Pain functions 2F84.

symptom or complaint back LS02
symptom or complaint low back LS03

LD67 Back syndrome with radiating pain

Description
Pain in the lumbar/thoracic region of the spine, accompanied by pain radiating to, or a neurological deficit of, an appropriate area; or sciatica, pain radiating down the

back of the leg, aggravated by coughing, movement, or posture; or demonstration of a prolapsed lumbar or thoracic disc by appropriate imaging technique, or during surgery.

Inclusion
disc prolapse/degeneration
lumbar disc prolapse with radiculopathy LD67.00
sciatica
thoracic disc prolapse with radiculopathy LD67.00

Exclusion
cervical disc lesion LD65
recent back strain LD66
spondylolisthesis LD66

Coding hint
For coding the problem level, consider Pain functions 2F84.

back pain LS02
low back pain LS03

Note
Exclude referred pain which is diffuse.

LD68 Shoulder syndrome

Description
Shoulder pain with limitation of movement/local tenderness/crepitus; or periarticular calcification on imaging.

Inclusion
adhesive capsulitis (frozen shoulder)
bursitis of shoulder
osteoarthrosis of shoulder
rotator cuff syndrome
synovitis of shoulder
tendinitis around shoulder

LD69 Patella disorder

Inclusion
recurrent instability of patella
retropatellar chondromalacia LD69.00

Exclusion
dislocation/subluxation due to an injury LD48

Coding hint
knee symptom LS14; sprain of knee LD45

LD70 Acquired deformity of spine

Inclusion
kyphoscoliosis
kyphosis
lordosis
scoliosis deformity of spine LD70.00

Exclusion
ankylosing spondylitis LD74
congenital deformity LD55
spondylolisthesis LD66

LD71 Acquired deformity of limb

Inclusion
acquired unequal limb length LD71.00
bunion
genu valgum-varum
hallux valgus/varus LD71.01
hammer toe LD71.02
mallet finger LD71.03
talipes (pes) planus (flatfoot) LD71.04

Exclusion
general congenital deformity/anomaly AD55
musculoskeletal genital deformity/anomaly LD55

LD72 Other specified and unknown bursitis, tendinitis, synovitis

Inclusion
acquired trigger finger LD72.00
bone spurs
bursitis LD72.01
calcaneus spur LD72.02
calcified tendon
Dupuytren's contracture LD72.03
fasciitis
ganglion
medial epicondylitis of elbow joint LD72.04
synovial cysts
tendinitis/tenosynovitis LD72.05
unknown bursitis, tendinitis, synovitis

Exclusion
bursitis/tendinitis/synovitis of shoulder LD68
sprain or strain of knee LD45
tennis elbow/lateral epicondylitis LD73

LD73 Tennis elbow

Description
A condition characterised by pain in or near the lateral humeral epicondyle or in the
forearm extensor muscle mass as result of unusual strain.

Inclusion
lateral epicondylitis

Exclusion
other tendinitis LD72

LD74 Rheumatoid arthritis and related conditions

Description

1. Rheumatoid arthritis (RA) is persistent and/or erosive disease that is defined as
   the confirmed presence of synovitis in at least one joint, absence of an alternative
   diagnosis that better explains the synovitis and achievement of a total score of 6 or
   greater (of a possible 10) from the individual scores in four domains: number and
   site of involved joints, serologic abnormality, elevated acute-phase response and
   symptom duration.
2. Adult onset Still's disease is a rare rheumatic condition characterised by a combination
   of symptoms, such as fever higher than 39 degrees C, cutaneous rash during fever
   peaks, joint or muscle pain, lymph node hypertrophy, increase of white blood cells
   (especially polymorphonuclear neutrophils) and abnormalities of liver metabolism.
3. Juvenile idiopathic arthritis (JIA) is the term used to describe a group of inflammatory articular disorders of unknown cause that begin before the age of 16 and last
   over 6 weeks. Six disorders have been defined: systemic-onset juvenile idiopathic
   arthritis (formerly referred to as Still's disease), oligoarticular arthritis, rheumatoid
   factor-positive polyarthritis, rheumatoid factor-negative polyarthritis, enthesitis-
   related arthritis (spondylarthropathies) and the juvenile form of psoriatic arthritis.
4. Ankylosing spondylitis is a chronic inflammatory condition affecting the axial
   joints, such as the sacroiliac joint and other intervertebral or costovertebral joints.
   It occurs predominantly in young males and is characterised by pain and stiffness
   of joints (ankylosis) with inflammation at tendon insertions.

Inclusion
adult-onset Still's disease
ankylosing spondylitis LD74.00
juvenile arthritis
rheumatoid arthritis LD74.01

Exclusion
gout LD75
other crystalarthropathies LD99
polymaylagia rheumatica LD75
psoriatic arthropathy LD99

Coding hint
For coding the problem level, consider Pain functions 2F84.

LD75 Gout

Description
Gout is an acute or chronic arthropathy resulting from deposition of monosodium urate monohydrate crystals in joint tissues. It is strongly associated with hyperuricaemia, which may be secondary to certain drugs, poisons or lymphoproliferative disorders. Gout is definitively diagnosed by demonstration of urate crystals in aspirated synovial fluid in the absence of an alternative aetiology for arthritis. It may be associated with focal urate deposition in skin and subcutaneous tissue (tophaceous gout) and with urate nephropathy.

Exclusion
hyperuricemia TD99
pseudo-gout/other crystal arthropathy LD99

Note
Gout is a term applied to a heterogeneous group of genetic and acquired diseases manifested by hyperuricemia and a characteristic acute inflammatory arthritis induced by crystals of monosodium urate monohydrate. Some patients develop aggregated deposits of these crystals (tophi) in and around the joints of the extremities that can lead to severe crippling. Many patients develop a chronic interstitial nephropathy. In addition, uric acid urolithiasis is common in gout. These manifestations of gout can occur in different combinations. However, essential hyperuricemia alone, even when complicated by uric acid lithiasis, should not be called gout; gout signifies inflammatory arthritis or tophaceous disease.

LD76 Polymyalgia rheumatica

Description
Polymyalgia rheumatica (PMR) is a syndrome characterised by aching of the proximal portions of the extremities and torso. Provisional classification criteria for PMR by the European League Against Rheumatism/American College of Rheumatology Collaborative Initiative should be applied to patients aged 50 years or older with bilateral shoulder aching and abnormal CRP and/or ESR. The scoring algorithm is based on morning stiffness >45 minutes (2 points), hip pain/limited range of motion (1 point), absence of rheumatoid factor and/or anti-citrullinated protein antibody (1 point), with optional ultrasound criteria. Most commonly, PMR occurs in isolation, but may be seen in 40–50% of patients with giant cell arteritis.

Coding hint
If also giant cell arteriitis (often together with polymyalgia), code KD99.

LD77 Osteochondrosis

Description
Any of a group of bone disorders involving one or more ossification centres (epiphyses). It is characterised by degeneration or necrosis followed by revascularisation and reossification. Osteochondrosis often occurs in children causing varying degrees of discomfort or pain. There are many eponymic types for specific affected areas, such as tarsal navicular (Kohler disease) and tibial tuberosity (Osgood-Schlatter disease).

Inclusion
apophysitis of calcaneus (Sevr's disease)
Legg-Calvé-Perthes disease LD77.00
Osgood-Schlatter disease LD77.01
osteochondritis dissecans LD77.02
Scheuermann's disease
slipped upper femoral epiphysis LD77.03
spinal endplate defects

Note
Osteochondroses are typically referred to by eponyms. The most common eponyms are indexed to osteochondrosis with specification identified by the site and time in life.

LD78 Osteoarthrosis of hip

Description
OA is the most common joint disease in persons 65 years of age and above. Its aetiology is not fully understood, although there are several related factors, such as female gender, genetics, metabolism and excessive mechanical stress. The diagnosis of OA is primarily based on clinical history and physical examination. The cardinal radiographic features of OA are focal/non-uniform narrowing of the joint space in the areas subjected to the most pressure, subchondral cysts, subchondral sclerosis and osteophytes. Osteoarthrosis means degeneration of the joint, and osteoarthritis means inflammation of the joint.

Inclusion
osteoarthritis of hip secondary to dysplasia/trauma

Coding hint
For coding the problem level, consider Pain functions 2F84.

arthritis NOS LD80
joint symptom LS20

LD79 Osteoarthrosis of knee

Description
A progressive, degenerative joint disease, the most common form of arthritis, especially in older persons. The disease is thought to result not from the ageing process but from biochemical changes and biomechanical stresses affecting articular cartilage. In the foreign literature, it is often called osteoarthrosis deformans.

Inclusion
osteoarthritis of knee secondary to dysplasia/trauma

Coding hint
For coding the problem level, consider Pain functions 2F84.

arthritis NOS LD80
joint symptom LS20

LD80 Other specified and unknown osteoarthrosis

Description
OA is the most common joint disease in persons 65 years of age and above. Its aetiology is not fully understood, although there are several related factors, such as female gender, genetics, metabolism and excessive mechanical stress. The diagnosis of OA is primarily based on clinical history and physical examination. The cardinal radiographic features of OA are focal/non-uniform narrowing of the joint space in the areas subjected to the most pressure, subchondral cysts, subchondral sclerosis and osteophytes. Osteoarthrosis means degeneration of the joint, and osteoarthritis means inflammation of the joint.

Inclusion
arthritis NOS
osteoarthritis
traumatic arthropathy
unknown osteoarthrosis

Exclusion
osteoarthosis of hip LD78
osteoarthrosis of knee LD79
osteoarthrosis of neck LD65
osteoarthrosis of shoulder LD68
osteoarthrosis of spine LD66

LD81 Osteoporosis

Description
Reduction of bone mass without alteration in the composition of bone, leading to fractures.

Inclusion
osteopenia LD81.00

Coding hint
For coding the problem level, consider Pain functions 2F84.

Note
Double code the pathological fracture due to osteoporosis.

LD99 Other specified diagnoses and diseases of musculoskeletal system

Inclusion
chronic internal derangement of knee
contractures
costochondritis
crystal arthropathy
dermatomyositis
hypermobility syndrome LD99.00
instability knee LD99.01
loose body in joint LD99.02
malunion of fracture
non-union of fracture (pseudoarthrosis) LD99.03
non-traumatic derangement of knee LD99.07
old meniscus injury LD99.04
osteomalacia
Paget's disease of bone
pathological fracture NOS
pseudo-gout
psoriatic arthritis LD99.05
Reiter's disease; scleroderma
Sjögren's syndrome
spontaneous rupture tendon
systemic lupus erythematosus
Tietze's disease LD99.06

Exclusion
gout LD75
hyperuricaemia TD99
post-polio paralysis ND01
post-stroke paralysis NS10

Coding hint
psoriatic arthritis (code also SD72)

N NEUROLOGICAL SYSTEM

NS SYMPTOMS, COMPLAINTS AND ABNORMAL FINDINGS OF NEUROLOGICAL SYSTEM

NS01 Headache

Inclusion
post-traumatic headache

Exclusion
atypical facial neuralgia ND99
cervicogenic headache LD65
cluster headache ND72
face pain NS02
migraine ND71
post-herpetic pain SD03
sinus pain RS11
tension headache ND73

Coding hint
For coding the problem level, consider Pain functions 2F84.

NS02 Pain, face

Exclusion
headache NS01
migraine ND71
post-herpetic pain SD03
sinus pain RS11
toothache DS19
trigeminal neuralgia ND74

Coding hint
For coding the problem level, consider Pain functions 2F84.

NS03 Restless legs

Description
A phenomenon characterised by aching or burning sensations in the lower and rarely the upper extremities that occur prior to sleep or may awaken the patient from sleep.

Inclusion
Sleep-related leg cramps

Exclusion
adverse effect medical agent AD40
intermittent claudication KD76
leg cramps LS13

Coding hint
For coding the problem level, consider Pain functions 2F84.
For coding the problem level, consider Sleep functions 2F72.

NS04 Tingling fingers, feet, toes

Inclusion
paraesthesia
prickly feeling fingers
prickly feeling feet
prickly feeling toes

Exclusion
pain or tenderness of skin SS01

NS05 Sensation disturbances

Inclusion
anaesthesia
burning sensation
numbness

Exclusion
pain/tenderness of skin SS01
tingling fingers, feet, toes NS04

NS06 Convulsion or seizure

Description
Clinical or subclinical disturbances of cortical function due to a sudden, abnormal, excessive and disorganised discharge of brain cells. Clinical manifestations include abnormal motor, sensory and psychic phenomena.

Inclusion
febrile convulsion
febrile seizures NS06.00
fit

Exclusion
fainting AS07
transient ischaemic attack ND68

NS07 Abnormal involuntary movements

Inclusion
dystonic movements
jerking
myoclonus
shaking
tetany
tremor
twitching

Exclusion
chorea KD02
convulsion NS06
cramps/spasm jaw LS06
cramps/spasm arm LS08
cramps/spasm hand/finger LS11
cramps/spasm leg/thigh LS13
cramps/spasm foot/toe LS16
cramps/spasm muscle LS17
dystonic disorder ND99
organic tic ND99
psychogenic tic ND99
restless legs NS03
tic douloureux ND74

NS08 Disturbance of smell, taste or both

Inclusion
anosmia

Exclusion
halitosis DS20

NS09 Vertigo or dizziness

Inclusion
giddiness
lightheaded NS09.00
loss of balance
rotatory vertigo NS09.01
woozy

Exclusion
motion sickness AD45
specific vertiginous syndrome HD67
syncope or blackout AS07

NS10 Paralysis and weakness

Inclusion
muscle weakness
palsy
paralysis
paralytic symptoms
paresis

Exclusion
general weakness AS04

NS11 Speech problem

Inclusion
aphasia
dysarthria
dysphasia
slurred speech
stammering or stuttering NS11.00

Exclusion
hoarseness RS13
speech delay PS18

NS90 Concern or fear of neurological disease

Description
Concern about/fear of neurological cancer in a patient without the disease, until the diagnosis is proven.

Coding hint
If the patient has the disease, code the disease.

NS99 Other specified symptoms, complaints and abnormal findings of neurological system

Inclusion
ataxia
gait abnormality
gait pattern problem
limping
meningism
transient global amnesia
walking problem

ND DIAGNOSES AND DISEASES OF NEUROLOGICAL SYSTEM
ND01 Poliomyelitis

Description
A disease of the nervous system, caused by human poliovirus. This disease commonly presents with a fever, sore throat, headache, vomiting or stiffness of the neck and back. This disease may present with an acute onset of flaccid paralysis. Transmission is commonly by the faecal-oral route or direct contact. Confirmation is by identification of poliovirus in a faecal sample or by a lumbar puncture.

Inclusion
acute poliomyelitis ND01.00
late effect of poliomyelitis
other neurological enterovirus infection
post-polio syndrome

ND02 Meningitis, encephalitis or both

Description
An acute febrile illness with abnormal findings in the cerebrospinal fluid.

Inclusion
bacterial meningitis ND02.00
encephalitis ND02.01
myelitis ND02.02
viral meningitis ND02.03

Coding hint
meningism NS99

ND03 Tetanus

Description
A disease of the skeletal muscle fibres, caused by an infection with the gram-positive bacteria **Clostridium tetani**. This disease is characterised by muscle spasms. Transmission is by direct contact of an open wound.

ND04 Other specified and unknown neurological infection

Inclusion
cerebral abscess
slow virus infection ND04.00
unknown neurological infection

Exclusion
acute polyneuritis ND01
meningitis/encephalitis ND02
poliomyelitis ND01

ND25 Neoplasm nervous system

Description
A benign, malignant or a neoplasm with uncertain behaviour that affects the brain, meninges, spinal cord, peripheral nerves or autonomic nervous system. Representative examples of primary neoplasms include astrocytoma, oligodendroglioma, ependymoma and meningioma.

Inclusion
benign neoplasm nervous system ND25.00
malignant neoplasm nervous system ND25.01
neoplasm of uncertain behaviour nervous system ND25.02

Exclusion
neurofibromatosis A90

Coding hint
unspecified neoplasm nervous system ND25.02

ND35 Concussion

Description
Concussion is a non-specific term used to describe transient alteration or loss of consciousness following closed head injury. The duration of unconsciousness generally lasts a few seconds, but may persist for several hours.

Inclusion
late effect of concussion

Exclusion
post-traumatic headache NS01

Coding hint
other head injury ND35
psychological effects of concussion PS02

ND36 Other specified and unknown head injury

Description
Trauma to the head, complicated by cerebral damage.

Inclusion
cerebral contusion ND36.00
cerebral injury with skull fracture
cerebral injury without skull fracture
epidural intracranial haematoma ND36.01
extradural haematoma
subdural haematoma

traumatic intracranial haemorrhage ND36.02
traumatic subdural intracranial haemorrhage ND36.03
unknown head injury

Exclusion
concussion ND35

ND37 Other specified and unknown injury neurological system

Inclusion
nerve injury
spinal cord injury
unknown injury neurological system

ND55 Congenital anomaly of neurological system

Inclusion
congenital hydrocephalus ND55.00
spina bifida ND55.01

ND65 Multiple sclerosis

Description
Multiple Sclerosis is a chronic, inflammatory demyelinating disease of the central nervous system. Three categories of multiple sclerosis have been outlined: relapsing/ remitting, secondary progressive and primary progressive multiple sclerosis. Multiple sclerosis is characterised by exacerbations/remissions of multiple neurological manifestation with deficits/derangements disseminated in both time and site (any combination of neurological signs and symptoms is possible).

Inclusion
disseminated sclerosis

Coding hint
For coding the problem level, consider Energy level 2F71.

ND66 Parkinsonism

Description
Parkinsonism is characterised by poverty and slowness of voluntary movements, resting tremor improving with active purposeful movement and muscular rigidity.

Parkinsonism is a clinical syndrome characterised by four cardinal features: rest tremor, muscular rigidity, akinesia or bradykinesia, and postural disturbances which include shuffling gait and flexed posture and loss of postural reflexes. Bradykinesia and one other clinical feature is required to make a diagnosis of Parkinsonism. Parkinsonism may result from a variety of conditions including progressive neurodegenerative disorders such as Parkinson's disease or atypical

Parkinsonism where the progressive degeneration of nigral and other neurons leads to dopamine deficiency. Parkinsonism may also be a result of structural lesions such as strokes or tumours or blockage of dopamine receptors in the striatum by drugs such as neuroleptics.

Inclusion
drug-induced Parkinsonism
paralysis agitans
Parkinson's disease ND66.00

ND67 Epilepsy

Description
Epilepsy is characterised by recurrent episodes of sudden altered consciousness, with or without tonic–clonic movements or seizure, plus either eyewitness account of the attack or characteristic abnormality of electroencephalogram (EEG).

Inclusion
focal seizures
generalised seizures
grand mal seizures
petit mal seizures
status epilepticus

Coding hint
convulsion NS05

ND68 Transient cerebral ischaemia

Description
Transient cerebral ischaemia is characterised by symptoms of transient (less than 24 hours) hypofunction of the brain, with sudden onset, presumed of vascular origin, without sequelae and with exclusion of migraine/migraine equivalent/epilepsy.

Inclusion
basilar insufficiency
drop attacks
transient ischaemic attack (TIA)

Exclusion
carotid bruit KS52
cerebrovascular accident ND69
migraine ND71
transient global amnesia NS99

Note
Double code with ND70.

ND69 Stroke or cerebrovascular accident

Description
Stroke is characterised by an acute neurological dysfunction caused by a focal infarction, presumed of vascular origin, lasting more than 24 hours or causing death, and within 4 weeks (28 days) of onset.

Inclusion
apoplexy
cerebral embolism
cerebral haemorrhage
cerebral infarction ND69.00
cerebral occlusion
cerebral stenosis
cerebral thrombosis
cerebrovascular accident
CVA
non-traumatic intracranial haemorrhage ND69.01
subarachnoid intracranial haemorrhage ND69.02

Exclusion
transient cerebral ischaemia ND68
traumatic intracranial haemorrhage ND36

Note
Double code with ND70.

ND70 Cerebrovascular disease

Description
This is a group of brain dysfunctions related to disease of the blood vessels supplying the brain. The criteria for this rubric is a previous transient cerebral ischaemia or stroke or investigation evidence of cerebrovascular disease.

Inclusion
cerebral aneurysm
sequelae of stroke

Exclusion
stroke or cerebrovascular accident ND69
transient cerebral ischaemia ND68

ND71 Migraine

Description
Migraine is characterised by recurrent episodes of headache with three or more of the following: unilateral headache; nausea/vomiting; aura; other neurological symptoms; family history of migraine.

Inclusion
vascular headache with aura
vascular headache without aura

Exclusion
cervicogenic headache LD65
cluster headache ND72
tension headache ND73

Coding hint
For coding the problem level, consider Pain functions 2F84.

ND72 Cluster headache

Description
Cluster headache is characterised by attacks of severe, often excruciating unilateral pain peri-orbitally and/or temporally, occurring up to eight times a day, sometimes associated with conjunctival injection, lacrimation, nasal congestion, rhinorrhoea, sweating, miosis, ptosis or eyelid oedema. Attacks occur in cluster periods lasting weeks or months, separated by remissions lasting months or years.

Coding hint
For coding the problem level, consider Pain functions 2F84.

ND73 Tension headache

Description
Tension headache is characterised by a pressing, generalised headache associated with stress and muscle tension with or without increased tenderness of pericranial muscles.

Exclusion
cluster headache ND72
migraine ND71

Coding hint
For coding the problem level, consider Pain functions 2F84.

ND74 Trigeminal neuralgia

Description
Trigeminal neuralgia is characterised by unilateral paroxysms of burning facial pain aggravated by touching trigger points, blowing nose or yawning, without sensory or motor paralysis.

Inclusion
tic douloureux

Exclusion
post-herpetic neuralgia SD03

Coding hint
For coding the problem level, consider Pain functions 2F84.
neuralgia NOS ND99

ND75 Facial paralysis

Description
Facial paralysis is characterised by an acute onset of unilateral paralysis of muscles of facial expression without sensory loss. Facial nerve dysfunction at the stylomastoid foramen leads to ipsilateral upper and lower facial weakness, manifested by an asymmetric smile, poor eyebrow elevation, decreased forehead wrinkling, widened palpebral fissure, weak eye closure, deviation of eye upward and laterally with attempted eye closure (Bell's phenomenon) and flattening of the nasolabial fold. Sagging of the lower eyelid causes tears to spill over the cheek, and saliva may also dribble from the corner of mouth. Although there may be subjective feelings of heaviness or numbness in the face, sensory loss is rarely demonstrable and taste is intact. If the lesion is in the middle ear portion proximal to the stylomastoid foramen, taste is lost over the anterior two-thirds of the tongue on same side. If the nerve to the stapedius is interrupted, there is hyperacusis (increased sensitivity to loud sounds).

Inclusion
Bell's palsy

ND76 Carpal tunnel syndrome

Description
Loss/impairment of superficial sensation affecting the thumb, index and middle finger, that may or may not split the ring finger. Dysaesthesia and pain worsen usually during the night and may radiate to the forearm.

ND77 Peripheral neuritis, neuropathy or both

Description
Sensory, reflex and motor changes confined to the territory of individual nerves, sometimes without apparent cause, sometimes secondary to a specific disease, e.g. diabetes.

Inclusion
acute infective polyneuropathy
common peroneal neuropathy ND77.00
diabetic neuropathy ND77.01
Guillain-Barré syndrome ND77.02
meralgia paresthetica ND77.03
Morton's neuroma ND77.04
neuritis

nerve lesion
phantom limb
phantom pain ND77.05
thoracic outlet syndrome ND77.06

Exclusion
post-herpetic neuropathy SD03

Note
Double code diabetic neuropathy with TD71, TD72.

ND99 Other specified and unknown diagnoses and diseases of neurological system

Inclusion
amyotrophic lateral sclerosis ND99.00
cerebral palsy
combined disorder of muscle and peripheral nerve ND99.01
motor neuron disease
myasthenia gravis ND99.02
neuralgia NOS
tic disorders ND99.03

Exclusion
sleep apnoea PS06

P PSYCHOLOGICAL, MENTAL AND NEURODEVELOPMENTAL

PS PSYCHOLOGICAL, MENTAL AND NEURODEVELOPMENTAL SYMPTOMS, COMPLAINTS AND ABNORMAL FINDINGS
PS01 Feeling anxious or nervous or tense

Description
Feelings of being anxious, nervous or tense, reported by the patient as an emotional or psychological experience not attributed to the presence of a mental disorder. A gradual transition exists from feelings that are unwelcome – but quite normal – and feelings that are so troublesome to the patient that professional help is sought.

Inclusion
anxiety NOS
feeling frightened

Exclusion
anxiety disorder PD06

PS02 Acute stress reaction

Description
A reaction to a stressful life event or significant life change requiring a major adjustment, either as an expected response to the event or as a maladaptive response interfering with daily coping and resulting in impaired social functioning, with recovery within a limited period of time.

Inclusion
acute adjustment problem
culture shock
feeling grief
feeling homesick
feeling stressed
immediate post-traumatic stress
shock (psychic)

Exclusion
depressive disorder PD12
feeling depressed PS03
post-traumatic stress disorder PD09

PS03 Feeling sad

Description
Feelings of sadness reported by the patient as an emotional or psychological experience not attributed to the presence of a mental disorder. A gradual transition exists from feelings that are unwelcome – but quite normal – and feelings that are so troublesome to the patient that professional help is sought.

Inclusion
feeling inadequate
unhappy

Exclusion
depressive disorder PD12
low self-esteem PS99

PS04 Feeling or being irritable or angry

Description
Feelings reported by the patient as an emotional or psychological experience not attributed to the presence of a mental disorder, or behaviour indicating irritability or anger. A gradual transition exists from feelings or behaviour that are unwelcome – but quite normal – and those that are so troublesome that professional help is sought.

Inclusion
agitation NOS
restlessness NOS

Exclusion
adolescent behaviour symptom/complaint PS19
child behaviour symptom/complaint PS18
irritability in partner ZC30
overactive child PS18

PS05 Suicidal ideation

Description
Thoughts, ideas or ruminations of thoughts about the possibility of ending one's life, ranging from thinking that one would be better off dead to formulation of elaborate plans.

Exclusion
suicide attempt PD14

PS06 Sleep disturbance

Description
Sleep disturbance as a diagnosis requires that the sleeping problem forms a major complaint, which, according to both patient and doctor, is not caused by another disorder but is a condition in its own right. Insomnia requires a quantitative or qualitative deficiency of sleep which is unsatisfactory in the patients' opinion, over a considerable period of time. In hypersomnia, excessive daytime sleepiness and sleep attacks exist which limit the patient's performance.

Inclusion
insomnia
nightmares
sleepwalking

Exclusion
jet lag AD45
sleep apnoea RS06
somnolence AS99

Coding hint
For coding the problem level, consider Sleep functions 2F72.

PS07 Sexual desire and fulfilment problem

Description
Sexual problems with regard to desire or to fulfilment not caused by any organic disorder or disease, but a reflection of the inability of a patient to participate in the sexual relationship she/he wants because of lack of desire, failure of genital response or function, or problems with sexual development.

Inclusion
frigidity
loss of libido
non-organic impotence or dyspareunia
premature ejaculation PS07.00
primary erectile dysfunction PS07.01
vaginismus of psychogenic origin PS07.02

Exclusion
concern with sexual preference PS08
organic impotence/sexual problems GS24
organic vaginismus GS23

Coding hint
For coding the problem level, consider Sexual functions 2F86.

PS08 Gender incongruence problem

Description
Gender incongruence is characterised by a marked and persistent incongruence between an individual's experienced gender and the assigned sex. Gender variant behaviour and preferences alone are not a basis for assigning the diagnoses in this group.

PS09 Eating problem in child

Description
Problem with eating behaviour of child.

Inclusion
feeding problem
food refusal

Exclusion
anorexia nervosa PD17
bulemia PD17
eating problem of infant/child TS04

Note
Problems with behaviour of children are particularly difficult to classify, which is illustrated by the fact that they are distributed over four chapters of the ICPC. Whether

or not parents present these problems to a GP will reflect their ideas about the gradual differences between normal – though maybe annoying – behaviour and behaviour that is considered worrying or 'pathological'.

PS10 Bedwetting or enuresis

Description
Bedwetting is characterised by involuntary voiding of urine by day/night not determined to be related to any organic disorder.

Exclusion
bedwetting due to organic disorder US03

Note
Problems with behaviour of children are particularly difficult to classify, which is illustrated by the fact that they are distributed over four chapters of the ICPC. Whether or not parents present these problems to a GP will reflect their ideas about the gradual differences between normal – though maybe annoying – behaviour and behaviour that is considered worrying or 'pathological'.

PS11 Encopresis

Description
Encopresis is the repeated production of usually well-formed faeces in inappropriate places like the floor or inside clothing, instead of on toilet or potty. In general the term 'encopresis' is used for children of at least 4 years of age and older, and not caused by constipation/sphincter control disorder/another disease.

PS12 Chronic alcohol problem

Description
A problem due to the use of alcohol resulting in one or more of the following: harmful use with clinically important damage to health; dependence syndrome; withdrawal state; psychotic disorder.

Inclusion
alcohol brain syndrome
alcohol dependence PS12.00
alcohol psychosis
alcohol withdrawal delirium PS12.01
alcoholism PS12.02
binge drinker PS12.03
delirium tremens
Korsakoff's psychosis PS12.04

Note

Substance abuse problem definitions should take into account the considerable differences between countries and cultures. A doctor can decide to label an episode as 'chronic alcohol abuse' without the patient's agreement and consequently also without the patient's willingness to any medical intervention.

PS13 Acute alcohol intoxication

Description

A problem due to the use of alcohol resulting in one or more of the following: acute intoxication; harmful use with clinically important damage to health; dependence syndrome; withdrawal state.

Inclusion

drunk

Note

A doctor can decide to label an episode as 'acute alcohol abuse' without the patient's agreement and consequently also without the patient's willingness to agree to any medical intervention.

PS14 Tobacco smoking problem

Description

A problem due to the use of tobacco resulting in one or more of the following: harmful use with clinically important damage to health; dependence syndrome; withdrawal state.

Inclusion

smoking problem

Note

Tobacco abuse/problem definitions should take into account the considerable differences between countries and cultures. An alcohol-dependent or heroin-addicted patient needs medical attention, but the definitions of 'tobacco abuse' are controversial. A physician can decide to label an episode as 'tobacco abuse' without the patient's agreement and consequently also without the patient's willingness to agree to any medical intervention.

PS15 Medication abuse

Description

Abuse of any prescribed medication.

Note

Substance abuse problem definitions should take into account the considerable differences between countries and cultures. Some patient's request and use tranquillizers, sleeping tablets, anorectics or laxatives inappropriately and for too long. In these cases physicians can decide to label the episode as 'medicine abuse' without the patients' agreement and consequently also without the patient's willingness to agree to any medical intervention.

PS16 Drug abuse

Description
A problem due to the use of a dependence-producing psychoactive substance, resulting in one or more of the following conditions: acute intoxication; harmful use with clinically important damage to health; dependence syndrome; withdrawal state; psychotic disorder.

Inclusion
addiction to drug
drug withdrawal
abuse or addiction hard drugs PS16.00
abuse or addiction soft drugs PS16.01

Note
Substance abuse problem definitions should take into account the considerable differences between countries and cultures. An alcohol-dependent or heroin-addicted patient needs medical attention, but the definitions of 'use of hashish' are controversial. Doctors can decide to label an episode as 'drug abuse' without the patient's agreement and consequently also without the patient's willingness to agree to any medical intervention.

PS17 Memory or attention problem

Inclusion
amnesia
disorientation
disturbance of concentration

PS18 Child behaviour symptom or complaint

Inclusion
delayed milestones
jealousy of child
overactive child
speech delay
temper tantrum

Exclusion
behaviour symptom/complaint adolescent PS19
behaviour symptom/complaint adult PD15
concern about physical development/growth delay TS08

PS19 Adolescent behaviour symptom or complaint

Inclusion
delinquency

Exclusion
behaviour symptom/complaint child PS18

PS20 Specific learning problems

Description
Specific speech, language and learning problems with onset in childhood, together with an impairment of functions related to biological maturation of the central nervous system, and a steady course over time without spontaneous remissions or relapses, although the deficit may diminish as the child grows older.

Inclusion
developmental disorder of motor function PS20.00
developmental language disorder PS20.01
developmental speech disorder PS20.02
dyslexia PS20.03

Exclusion
attention deficit disorder PD16
mental retardation PD18

PS21 Own illness problem

Inclusion
dependence on others PS21.00
problems related to adherence to medical advice

PS22 Phase of life problem

Inclusion
empty-nest problem PS22.00
feeling old
old age
retirement problem PS22.01
senescence

PS90 Concern, fear of mental disorder or problem

Description
Concern about/fear of mental disease in a patient without the disease, until the diagnosis is proven.

Inclusion
concern about mental disease
fear of committing suicide

Coding hint
If the patient has the disease, code the disease.

PS99 Other specified psychological/mental symptom/ complaint/abnormal finding

Inclusion
delusions
eating disorders NOS
hallucinations
hyperactivity
multiple psychological symptoms/complaints
poor hygiene
strange behaviour
suspiciousness

PD PSYCHOLOGICAL, MENTAL AND NEURODEVELOPMENTAL DIAGNOSES AND DISEASES
PD01 Dementia

Description
Dementia is a syndrome due to a disease of the brain, usually of a chronic and/or progressive nature, with clinically significant disturbance of multiple higher cortical functions (memory, thinking, orientation, comprehension), together with intact consciousness.

Inclusion
Alzheimer's disease PD01.00
multi-infarct dementia PD01.01
senile dementia

Coding hint
other psychological symptoms PS99
phase of life problem ZC02

PD02 Other specified and unknown organic mental disorder

Description
Organic mental disorders as a diagnosis require psychological syndromes, patterns or behaviour due to organic disease.

Inclusion
delirium PD02.00
unknown organic mental disorder

Exclusion
psychosis caused by alcohol PS12
other specified psychosis PD05

PD03 Schizophrenia

Description
Schizophrenia is characterised by disturbances in multiple mental modalities, including thinking (e.g. delusions, disorganisation in the form of thought), perception (e.g. hallucinations), self-experience (e.g. the experience that one's feelings, impulses, thoughts or behaviour are under the control of an external force), cognition (e.g. impaired attention, verbal memory and social cognition), volition (e.g. loss of motivation), affect (e.g. blunted emotional expression) and behaviour (e.g. behaviour that appears bizarre or purposeless, unpredictable or inappropriate emotional responses that interfere with the organisation of behaviour). Psychomotor disturbances, including catatonia, may be present. Persistent delusions, persistent hallucinations, thought disorder and experiences of influence, passivity or control are considered core symptoms. Symptoms must have persisted for at least 1 month in order for a diagnosis of schizophrenia to be assigned. The symptoms are not a manifestation of another health condition (e.g. a brain tumour) and are not due to the effect of a substance or medication on the central nervous system (e.g. corticosteroids), including withdrawal (e.g. alcohol withdrawal).

Inclusion
all types of paranoia
all types of schizophrenia

Exclusion
acute/transient psychosis PD05

PD04 Affective psychosis

Description
A fundamental disturbance in affect and mood (with/without associated anxiety). In manic disorder mood, energy and activity are simultaneously elevated. In bipolar disease, at least two periods of disturbed mood, shifting from elevated to lowered are observed.

Inclusion
bipolar disorder PD04.00
hypomania
mania
manic depression

Exclusion
depression PD12

Coding hint
psychosis NOS PD05

PD05 Other specified or unknown psychosis

Inclusion
acute psychosis
puerperal psychosis
reactive psychosis
transient psychosis
unknown psychosis

PD06 Anxiety disorder or anxiety state

Description
Clinically significant anxiety that is not restricted to any particular environmental situation. It manifests as a panic disorder (recurrent attacks of severe anxiety not restricted to any particular situation, with or without physical symptoms) or as a disorder in which generalised and persistent anxiety, not related to any particular situation, occurs with variable physical symptoms.

Inclusion
generalised anxiety disorder PD06.00
panic disorder PD06.01
phobia PD06.02

Exclusion
anxiety NOS PS01
anxiety with depression PD12

Coding hint
feeling anxious/nervous/tense PS01

PD07 Obsessive-compulsive or related disorder

Description
Obsessive-compulsive and related disorders is a group of disorders characterised by repetitive thoughts and behaviours that are believed to share similarities in aetiology and key diagnostic validators. Cognitive phenomena such as obsessions, intrusive thoughts and preoccupations are central to a subset of these conditions (i.e. obsessive-compulsive disorder, body dysmorphic disorder, hypochondriasis and olfactory reference disorder) and are accompanied by related repetitive behaviours. Hoarding disorder is not associated with intrusive unwanted thoughts but rather is characterised by a compulsive need to accumulate possessions and distress related to

discarding them. Also included in the grouping are body-focused repetitive behaviour disorders, which are primarily characterised by recurrent and habitual actions directed at the integument (e.g. hair-pulling, skin-picking) and lack a prominent cognitive aspect. The symptoms result in significant distress or significant impairment in personal, family, social, educational, occupational or other important areas of functioning.

Inclusion
hoarding disorder
hypochondriasis

PD08 Adjustment disorders

Description
Disorders specifically associated with stress are directly related to exposure to a stressful or traumatic event, or a series of such events or adverse experiences. For each of the disorders in this grouping, an identifiable stressor is a necessary, though not sufficient, causal factor. Although not all individuals exposed to an identified stressor will develop a disorder, the disorders in this grouping would not have occurred without experiencing the stressor. Stressful events for some disorders in this grouping are within the normal range of life experiences (e.g. divorce, socio-economic problems, bereavement). Other disorders require the experience of a stressor of an extremely threatening or horrific nature (i.e. potentially traumatic events). With all disorders in this grouping, it is the nature, pattern and duration of the symptoms that arise in response to the stressful events together with associated functional impairment that distinguishes the disorders.

Inclusion
persistent adjustment disorder
prolonged grief disorder

Exclusion
acute stress reaction PS02

PD09 Post-traumatic stress disorder

Description
A stressful event followed by a major state of distress and disturbance, with a delayed or protracted reaction, flashbacks, nightmares, emotional blunting and anhedonia interfering with social functioning and performance, and including depressed mood, anxiety, worry and feeling unable to cope, persistent over time.

   Complex post-traumatic stress disorder (Complex PTSD) is a disorder that may develop following exposure to an event or series of events of an extremely threatening or horrific nature, most commonly prolonged or repetitive events from which escape is difficult or impossible (e.g. torture, slavery, genocide campaigns, prolonged domestic violence, repeated childhood sexual or physical abuse). All diagnostic requirements for PTSD are met.

In addition, Complex PTSD is characterised by severe and persistent problems in affect regulation; beliefs about oneself as diminished, defeated or worthless, accompanied by feelings of shame, guilt or failure related to the traumatic event; and difficulties in sustaining relationships and in feeling close to others. These symptoms cause significant impairment in personal, family, social, educational, occupational or other important areas of functioning.

Inclusion
complex post-traumatic stress syndrome

Coding hint
feeling anxious PS01
acute stress reaction PS02
feeling depressed PS03
enduring personality change after catastrophic experience PD15
For coding the problem level, consider Sleep function 2F72.

PD10 Bodily distress or somatisation disorder

Description
Bodily distress disorder is characterised by the presence of bodily symptoms that are distressing to the individual and excessive attention directed toward the symptoms, which may be manifest by repeated contact with health care providers. If another health condition is causing or contributing to the symptoms, the degree of attention is clearly excessive in relation to its nature and progression. Excessive attention is not alleviated by appropriate clinical examination and investigations and appropriate reassurance. Bodily symptoms are persistent, being present on most days for at least several months. Typically, bodily distress disorder involves multiple bodily symptoms that may vary over time. Occasionally there is a single symptom, usually pain or fatigue, that is associated with the other features of the disorder.

Inclusion
somatisation disorder

Note
Consider using a symptom diagnosis instead of labelling bodily distress as a disorder. Bodily distress is linked to the 'old' concepts of somatisation and somatoform disorders and to new 'concepts' as 'somatic symptom disorder or medically unexplained symptoms'. In primary care the use of symptom diagnoses without a 'psychogenic' connotation is advised instead of somatic symptom disorder or medically unexplained symptoms.

PD11 Burn-out

Description

Burn-out is a syndrome conceptualised as resulting from chronic workplace stress that has not been successfully managed. It is characterised by three dimensions: a) feelings of energy depletion or exhaustion; b) increased mental distance from one's job or feelings of negativism or cynicism related to one's job; and c) reduced professional efficacy. Burn-out refers specifically to phenomena in the occupational context and should not be applied to describe experiences in other areas of life.

Because burn-out is not only related to workplace stress, the description for neurasthenia is also presented here: burn-out is characterised by increased fatiguability with unpleasant associations, difficulties in concentration and a persistent decrease in performance and coping efficiency; the feeling of physical weakness and exhaustion after mental effort or after a minimal physical effort is often accompanied by muscular pain and an inability to relax.

Inclusion

neurasthenia
surmenage

Exclusion

chronic fatigue syndrome AS04

Coding hint

For coding the problem level, consider Sleep functions 2F72 and Energy level 2F71.

PD12 Depressive disorder

Description

A depressive disorder is characterised by fundamental disturbance in affect and mood towards depression (continuum with feeling sad/depressed and diagnosis depression). Mood, energy and activity are simultaneously lowered, together with an impaired capacity for enjoyment, interest and concentration. Sleep and appetite are usually disturbed and self-esteem and confidence are decreased.

Inclusion

depressive psychosis
dysthymia PD12.00
postpartum depression PD12.01
puerperal depression
reactive depression

Exclusion

feeling sad PS03
mixed anxiety depression disorder PD13

Coding hint
For coding the problem level, consider Sleep functions 2F72 and Energy level 2F71.

PD13 Mixed depressive and anxiety disorder

Description
Mixed depressive and anxiety disorder is characterised by symptoms of both anxiety and depression more days than not for a period of 2 weeks or more. Neither set of symptoms, considered separately, is sufficiently severe, numerous or persistent to justify a diagnosis of a depressive episode, dysthymia or an anxiety- and fear-related disorder. Depressed mood or diminished interest in activities must be present accompanied by additional depressive symptoms as well as multiple symptoms of anxiety. The symptoms result in significant distress or significant impairment in personal, family, social, educational, occupational or other important areas of functioning. There have never been any prior manic, hypomanic or mixed episodes, which would indicate the presence of a bipolar disorder.

Inclusion
mixed anxiety and depression

Exclusion
depressive disorder PD12

Coding hint
For coding the problem level, consider Sleep functions 2F72 and Energy level 2F71.

PD14 Suicide or suicide attempt

Description
A successful ending of one's life or self-harming behaviour undertaken with the intention of ending one's life.

Inclusion
successful attempt
suicide PD14.01
suicide attempt PD14.00
suicide gesture

Exclusion
afraid of committing suicide PS90
suicidal ideation PS05

Note
In case of suicide, double code with AD96.

PD15 Personality disorder

Description

Personality disorder is characterised by problems in functioning of aspects of the self (e.g. identity, self-worth, accuracy of self-view, self-direction) and/or interpersonal dysfunction (e.g. ability to develop and maintain close and mutually satisfying relationships, ability to understand others' perspectives and to manage conflict in relationships) that have persisted over an extended period of time (e.g. 2 years or more). The disturbance is manifest in patterns of cognition, emotional experience, emotional expression and behaviour that are maladaptive (e.g. inflexible or poorly regulated) and is manifest across a range of personal and social situations (i.e. is not limited to specific relationships or social roles). The patterns of behaviour characterising the disturbance are not developmentally appropriate and cannot be explained primarily by social or cultural factors, including socio political conflict. The disturbance is associated with substantial distress or significant impairment in personal, family, social, educational, occupational or other important areas of functioning.

Inclusion

adult behaviour disorder
borderline personality disorder PD15.00

PD16 Attention deficit hyperactivity disorder

Description

Attention deficit hyperactivity disorder is characterised by a persistent pattern (at least 6 months) of inattention and/or hyperactivity-impulsivity, with onset during the developmental period, typically early to mid childhood. The degree of inattention and hyperactivity-impulsivity is outside the limits of normal variation expected for age and level of intellectual functioning and significantly interferes with academic, occupational or social functioning. Inattention refers to significant difficulty in sustaining attention to tasks that do not provide a high level of stimulation or frequent rewards, distractibility and problems with organisation. Hyperactivity refers to excessive motor activity and difficulties with remaining still, most evident in structured situations that require behavioural self-control. Impulsivity is a tendency to act in response to immediate stimuli, without deliberation or consideration of the risks and consequences. The relative balance and the specific manifestations of inattentive and hyperactive-impulsive characteristics varies across individuals and may change over the course of development. In order for a diagnosis of disorder, the behaviour pattern must be clearly observable in more than one setting.

Inclusion

attention deficit disorder (ADD)
hyperkinetic disorder

Exclusion

adolescent behaviour symptom/complaint PS19
learning disorder PS20

PD17 Eating disorder

Description
Eating disorders involve abnormal eating or feeding behaviours that are not explained by another health condition and are not developmentally appropriate or culturally sanctioned. Feeding disorders involve behavioural disturbances that are not related to body weight and shape concerns, such as eating of non-edible substances or voluntary regurgitation of foods. Eating disorders include abnormal eating behaviour and preoccupation with food as well as prominent body weight and shape concerns.

Inclusion
anorexia nervosa PD17.00
bulimia PD17.01
binge eating
pica

Coding hint
eating problem in child, food refusal PS09
feeding problem infant/child TS04
feeding problem adult TS05

PD18 Disorders of intellectual development

Description
Arrested/incomplete development of the mind with impairment of skills during the developmental period and a low overall level of intelligence, with/without impairment of behaviour.

Exclusion
mental retardation due to congenital anomaly AD55

PD19 Autism spectrum disorders

Description
Autism spectrum disorder is characterised by persistent deficits in the ability to initiate and to sustain reciprocal social interaction and social communication, and by a range of restricted, repetitive and inflexible patterns of behaviour and interests. The onset of the disorder occurs during the developmental period, typically in early childhood, but symptoms may not become fully manifest until later, when social demands exceed limited capacities. Deficits are sufficiently severe to cause impairment in personal, family, social, educational, occupational or other important areas of functioning and are usually a pervasive feature of the individual's functioning observable in all settings, although they may vary according to social, educational or other context. Individuals along the spectrum exhibit a full range of intellectual functioning and language abilities.

Inclusion
Asperger syndrome
autistic disorder PD19.00

PD99 Other specified or unknown psychological or mental diagnoses or diseases

Inclusion
compulsive gambling PD99.00
Munchausen's syndrome
neurosis

R RESPIRATORY SYSTEM

RS SYMPTOMS, COMPLAINTS AND ABNORMAL FINDINGS OF RESPIRATORY SYSTEM
RS01 Pain respiratory system

Inclusion
painful respiration
pleuritic pain
pleurodynia

Exclusion
chest pain AS12
musculoskeletal chest pain LS04
nose pain RS10
sinus pain RS11
sore throat RS12
chest tightness RS99
pleurisy RS50

RS02 Shortness of breath

Inclusion
orthopnoea

Exclusion
hyperventilation RS04
stridor RS04
wheezing RS03

RS03 Wheezing

Description
Continuous adventitious sounds that are high-pitched are called wheezes. Wheezes originate in airways narrowed by spasm, thickening of the mucosa or luminal obstruction.

Inclusion
expiratory wheeze
rhonchi

Exclusion
dyspnoea RS02
hyperventilation RS04
stridor RS04

RS04 Other specified breathing problem

Inclusion
abnormal breathing
apnoea
holding breath
hyperventilation
inspiratory wheeze
respiratory distress
stridor
tachypnoea

RS05 Snoring

RS06 Sleep-related breathing problems

Inclusion
central sleep apnoea
obstructive sleep apnoea
sleep apnoea RS06.00

RS07 Cough

Description
Cough is an important natural defensive mechanism and protective reflex for clearing the upper and lower airways of excessive secretions such as mucus and inhaled particles. Cough is a common symptom of most respiratory disorders and may be indicative of trivial to very serious airway or lung pathology.

Inclusion
dry cough
moist cough

Exclusion
abnormal sputum/phlegm RS15

RS08 Nose bleed or epistaxis

RS09 Sneezing or nasal congestion

Inclusion
blocked nose
rhinorrhea
running nose

RS10 Nose symptoms or complaints

Inclusion
pain in nose

Exclusion
anosmia NS08
blocked nose/sneezing RS09
complaint of sinuses RS11
concern with appearance of nose RS91
epistaxis RS08
rhinophyma SD99

RS11 Sinus symptoms or complaints

Inclusion
blocked sinus
congested sinus
pain/pressure in sinus
post-nasal drip

Exclusion
headache NS01
face pain NS02
nasal congestion RS09

RS12 Throat symptoms or complaints

Inclusion
dry throat
inflamed throat
red throat
sore throat
large tonsils

lump in throat
pain in throat RS12.00
tonsillar pain

Exclusion
tonsillar hypertrophy RD66
voice symptom RS13

RS13 Voice symptoms or complaints

Inclusion
absence of voice
aphonia
hoarseness

Exclusion
neurological disorder of speech NS11
stammering/stuttering NS11
sore throat RS12

RS14 Haemoptysis or coughing blood

RS15 Abnormal sputum or phlegm

Exclusion
cough with sputum RS07
haemoptysis RS14

RS50 Pleurisy or pleural effusion

Description
To classify pleurisy/pleural effusion, there should be clinical evidence of pleural exudate or pleuritic pain accompanied by pleural friction rub or investigative evidence of inflammatory pleural exudate.

Inclusion
pleural inflammatory exudate
pleuritis

Exclusion
pneumonia RD09
tuberculosis AD15

Coding hint
pleuritic pain RS01

Note
malignant effusion to be coded to origin of malignancy

RS90 Concern or fear of disease respiratory system

Description
Concern about/fear of disease in a patient without the disease, until the diagnosis is proven.

Note
If patient has the disease, code the disease.

RS91 Concern about appearance of nose

Inclusion
red nose
prominent nose

RS99 Other specified respiratory symptoms, complaints and abnormal findings

Inclusion
chest tightness
fluid on lung
hiccough RS99.00
irritable airways RS99.01
lung congestion

RD DIAGNOSES AND DISEASES OF RESPIRATORY SYSTEM
RD01 Pertussis

Description
A disease of the upper respiratory tract, caused by an infection of the Gram-negative bacteria **Bordetella pertussis**. This disease typically presents with paroxysmal cough, inspiratory whoop, and fainting or vomiting after coughing. Transmission is by inhalation of infected respiratory secretions.

Inclusion
parapertussis

RD02 Acute upper respiratory infection

Description
Upper respiratory infection (URI) is characterised by evidence of acute inflammation of nasal or pharyngeal mucosa with absence of criteria for more specifically defined acute respiratory infection classified in this section.

Inclusion
acute pharyngitis RD02.00
acute rhinitis
common cold RD02.01
coryza
head cold
nasopharyngitis
pharyngitis
URI
URTI

Exclusion
allergic rhinitis RD65
chronic pharyngitis RD10
infectious mononucleosis AD04
influenza RD07
laryngitis/croup RD05
measles AD01
sinusitis RD03
tonsillitis/quinsy RD04
viral pharyngoconjunctivitis FD01

RD03 Acute or chronic rhinosinusitis

Description
Rhinosinusitis is characterised by purulent nasal/post-nasal discharge, or previous medically treated episodes of sinusitis, plus tenderness over one/more sinuses, or deep-seated aching facial pain aggravated by dependency of head, or opacity on trans-illumination; or imaging evidence of sinusitis; or pus obtained from the sinus.

Inclusion
acute sinusitis RD03.00
chronic sinusitis RD03.01
sinusitis affecting any paranasal sinus

Coding hint
face pain NS02
headache NS01
upper respiratory tract infection RD02

RD04 Acute tonsillitis

Description
Acute tonsillitis is characterised by sore throat or fever with reddening of tonsil(s) more than the posterior pharyngeal wall, and either pus on swollen tonsil(s) or enlarged tender regional lymph node. Strep throat is an acute inflammation of the throat, plus demonstration of beta-haemolytic streptococci.

Inclusion
peritonsillar abscess RD04.00
streptococcal throat RD04.01

Exclusion
diphtheria RD10
erysipelas/strep skin infection SD16
hypertrophy/chronic infection of tonsils RD66
infectious mononucleosis AD04
scarlet fever AD24

RD05 Acute (obstructive) laryngitis or tracheitis or both

Description
Acute laryngitis and tracheitis are defined respectively as acute inflammation of larynx and trachea, with local findings of erythema and oedema of laryngeal and tracheal mucosa. Acute laryngitis and tracheitis are induced by upper respiratory tract viral infections or voice abuse. Acute obstructive laryngitis (croup) is a condition commonly caused by an acute viral infection of the upper airway. This condition is characterised by a barking cough, stridor, hoarseness or difficulty breathing. Transmission is commonly by inhalation of infected respiratory secretions.

Inclusion
acute subglottis laryngitis RD05.00
croup

Exclusion
epiglottitis RD10
false croup/pseudocroup (laryngeal spasm) RD99

Coding hint
upper respiratory tract infection RD02

RD06 Acute bronchitis or bronchiolitis or both

Description
An acute disease of the bronchi, commonly caused by an infection with a bacterial or viral source. This disease is characterised by inflammation of the bronchi. This disease presents with cough, wheezing, chest pain or discomfort, fever or dyspnoea. Transmission is by inhalation of infected respiratory secretions. Bronchiolitis is an acute disease of the bronchioles, commonly caused by an infection with a bacteria or viral source. This disease is characterised by inflammation of the bronchioles and coryza. This disease presents with cough, wheezing, tachypnoea, fever or chest retraction. In children and adults the disease is characterised by cough and fever with scattered or generalised abnormal chest signs: wheeze, coarse rales, rhonchi or moist sounds; in infants (bronchiolitis): dyspnoea and hyperinflation.

Inclusion
acute lower respiratory infection NOS
bronchitis NOS
chest infection NOS
tracheobronchitis

Exclusion
allergic bronchitis RD69
chronic bronchitis RD67
influenza RD07

Coding hint
cough RS07
upper respiratory tract infection RD02
wheezing RS03

RD07 Influenza

Description
Influenza is characterised by myalgia and cough without abnormal respiratory physical signs other than inflammation of nasal mucous membrane and throat, plus three or more of the following: sudden onset (within 12 hours); rigors/chills/fever; prostration and weakness; influenza in close contacts; influenza epidemic; or viral culture/serological evidence of influenza virus infection.

Inclusion
influenza-like illness
para-influenza

Exclusion
gastric flu DD05
influenza pneumonia RD09

Coding hint
fever AS03
upper respiratory tract infection RD02
virus infection NOS AD14

RD08 Coronavirus disease 2019 (COVID-19)

Coding hint
For coding the problem level, consider Energy level 2F71.

RD09 Pneumonia

Description
A disease of the lungs, frequently but not always caused by an infection with bacteria, virus, fungus or parasite. This disease is characterised by fever, chills, cough with sputum production, chest pain and shortness of breath.

Inclusion
bacterial pneumonia
bronchopneumonia
influenzal pneumonia
legionella pneumonia RD09.00
viral pneumonia

Exclusion
aspiration pneumonia RD99

Coding hint
acute bronchitis RD06
cough RS07

RD10 Other specified or unknown respiratory infection

Inclusion
chronic nasopharyngitis
chronic pharyngitis
chronic rhinitis NOS
diphtheria RD10.00
empyema
epiglottitis RD10.01
fungal respiratory infection
lung abscess
protozoal infection (without pneumonia)
severe acute respiratory syndrome (SARS) RD10.02
unknown respiratory infection

RD25 Malignant neoplasm bronchus and lung

Description
A characteristic histological appearance of a primary or metastatic malignant neoplasm in the lung.

Inclusion
malignancy of bronchus
malignancy of lung
malignancy of trachea

Coding hint
uncertain or carcinoma in situ respiratory neoplasm RD28

RD26 Other specified or unknown respiratory malignant neoplasm

Description
Characteristic histological appearance.

Inclusion
malignancy of larynx
mediastinum
nose
pharynx
pleura
sinus; mesothelioma
unknown respiratory malignant neoplasm

Exclusion
Hodgkin's disease BD25
malignancy of trachea/bronchus/lung RD25

Coding hint
unspecified respiratory neoplasm RD28

RD27 Benign neoplasm respiratory

Description
Characteristic clinical or histological appearance.

Exclusion
nasal polyp RD99
unspecified respiratory neoplasm RD28

RD28 Uncertain or carcinoma in situ neoplasm of respiratory system

Inclusion
respiratory neoplasm not specified as benign or malignant when histology is not available

Exclusion
benign respiratory neoplasm RD27
malignant neoplasm bronchus/lung RD25
malignant neoplasm respiratory, other RD26
secondary neoplasm unknown site AD25

RD35 Injury respiratory system

Inclusion
trauma to nose
trauma to respiratory system

Exclusion
drowning AD45
fractured nose LD39
foreign body in respiratory system RD36

Coding hint
In case of pneumothorax due to injury, code also RD99.04.

RD36 Foreign body in nose, larynx, bronchus

Inclusion
foreign body in lung

Exclusion
aspiration pneumonia RD99
drowning AD45
foreign body lodged in oesophagus DD36
foreign body in ear HD36

Coding hint
other complaint of respiratory system RS99

RD55 Congenital anomaly of respiratory system

Inclusion
congenital abnormality of bronchi
congenital abnormality of larynx
congenital abnormality of lungs
congenital abnormality of nose
congenital abnormality of pharynx
congenital abnormality of pleura
congenital abnormality of trachea

Exclusion
cleft lip/palate DD55
cystic fibrosis TD99

RD65 Allergic rhinitis

Description
Rhinitis is inflammation of the nasal mucosa clinically characterised by major symptoms: sneezing, nasal pruritus, running nose and stuffy nose. Allergic rhinitis is

an inflammation of nasal airway triggered by allergens to which the affected individual has previously been sensitised.

Inclusion
allergic
hay fever
nasal allergy
pollen
seasonal
vasomotor rhinitis

Exclusion
chronic rhinitis NOS RD10
upper respiratory tract infection RD02

RD66 Hypertrophy tonsils or adenoids or both

Description
Any persistent or recurrent disease affecting the round-to-oval mass of lymphoid tissue embedded in the lateral wall of the pharynx (tonsils) or the collection of lymphoid nodules on the posterior wall and roof of the nasopharynx (adenoids) resulting in enlargement of the tonsils or adenoids or both. (ICD-11)

Inclusion
chronic tonsillitis

Exclusion
acute tonsillitis RD04
allergic rhinitis RD65

RD67 Chronic bronchitis

Description
Chronic bronchitis is an unspecified chronic inflammation of the bronchi (medium-size airways) in the lungs, causing a persistent cough that produces sputum (phlegm) and mucus for at least 3 months per year in 2 consecutive years.

Exclusion
emphysema/chronic obstructive pulmonary (lung, airways) disease RD68
bronchiectasis RD99

Coding hint
abnormal sputum/phlegm RS15
acute bronchitis RD06
cough RS07

RD68 Chronic obstructive pulmonary disease and emphysema

Description

Chronic obstructive pulmonary disease (COPD), a common preventable and treatable disease, is characterised by persistent airflow limitation that is usually progressive and associated with an enhanced chronic inflammatory response in the airways and the lung to noxious particles or gases. Exacerbations and comorbidities contribute to the overall severity in individual patients. Emphysema is defined by abnormal and permanent enlargement of the airspaces that are distal to the terminal bronchioles. This is accompanied by destruction of the airspace walls, without obvious fibrosis (i.e. there is no fibrosis visible to the naked eye). Emphysema can exist in individuals who do not have airflow obstruction; however, it is more common among patients who have moderate or severe airflow obstruction.

Inclusion

chronic airways limitation (CAL)
chronic obstructive airways disease (COAD)
chronic obstructive lung disease (COLD)
chronic obstructive pulmonary disease (COPD)
emphysema

Exclusion

asthma RD69
bronchiectasis RD99
chronic bronchitis RD67
cystic fibrosis RD99

Coding hint

For coding the problem level, consider Energy level 2F71.

RD69 Asthma

Description

Asthma is characterised by recurrent episodes of reversible acute bronchial obstruction with wheeze/dry cough or diagnostic test meeting currently accepted criteria for asthma.

Inclusion

allergic asthma RD69.00
reactive airways disease
wheezy bronchitis

Exclusion

bronchiolitis RD06
chronic bronchitis RD67
emphysema RD68

Coding hint
cough RS07
wheezing RS03

RD70 Lung disease related to external agents

Inclusion
pneumoconiosis RD70.00
pneumonitis due to allergy
pneumonitis due to chemicals
pneumonitis due to dust
pneumonitis due to fumes
pneumonitis due to mould
vaping-related disorder RD70.01

Exclusion
air pollution AD45

RD99 Other specified or unknown diagnoses and diseases of respiratory system

Inclusion
aspiration pneumonia RD99.00
bronchiectasis RD99.01
cystic fibrosis RD99.02
deviated nasal septum RD99.03
interstitial lung disease RD99.08
lung complication of other disease
mediastinal disease
other disease of larynx
pneumothorax RD99.04
polyp of nasal cavity RD99.05
polyp of vocal cord RD99.06
pulmonary collapse
pulmonary oedema without heart disease/heart failure
respiratory failure RD99.07

S SKIN

SS SYMPTOMS, COMPLAINTS AND ABNORMAL FINDINGS OF SKIN

SS01 Pain or tenderness of skin

Inclusion
painful lesion or rash
soreness

Exclusion
other sensation disturbance NS05
tingling fingers, feet, toes NS04

SS02 Pruritus

Inclusion
skin irritation

Exclusion
dermatitis artefacta SD99
nipple pruritus GS27
perianal itching DS05
scrotal pruritus GS21
vulval pruritus GS18

SS03 Lump or swelling of skin localised

Inclusion
papule

Exclusion
breast lump GS26
insect bite SD39
swelling AS09

SS04 Lump or swelling of skin generalised

Inclusion
lumps in multiple sites
papules in multiple sites
swellings in multiple sites

Exclusion
ankle oedema KS04
swelling AS09

SS05 Rash localised

Inclusion
blotch
erythema
redness

Exclusion
localised lump or swelling of skin SS03

SS06 Rash generalised

Inclusion
blotches occurring in multiple sites
erythema occurring in multiple sites
redness occurring in multiple sites

Exclusion
generalised lumps/swellings skin SS04
other viral exanthem AD13

SS07 Skin colour change

Inclusion
circles under eyes
cyanosis SS07.00
flushing
freckles
livedo reticularis
pallor

Exclusion
bruise SD35
hot flushes GS13
jaundice DS13
vitiligo SD99

SS08 Skin texture symptom or complaint

Inclusion
dry skin SS08.00
induration of skin SS08.01
fissura skin
peeling
scaling
wrinkles

Exclusion
ichthyosis SD55
scalp symptom/complaint SS11
sweating problem AS10
sweat gland disease SD73
vulval symptom/complaint GS18

SS09 Nail symptom or complaint

Inclusion
clubbing

Exclusion
ingrowing nail SD75
paronychia SD05

SS10 Hair loss or baldness

SS11 Other specified hair or scalp symptom or complaint

Inclusion
change in hair colour SS11.00
dry scalp
hirsutism SS11.01

Exclusion
dandruff SD68
folliculitis SD06
hair loss/baldness SS10
trichotillomania PS99

SS90 Concern or fear of disease of skin

Description
Concern about/fear of disease of skin in a patient without the disease, until the diagnosis is proven.

Coding hint
If the patient has the disease, code the disease.

SS99 Other specified symptoms, complaints and abnormal findings of skin

Inclusion
cellulite
petechiae
problems with umbilicus
sore(s)
spontaneous ecchymosis

Exclusion
chronic ulcer skin/pressure sore SD77
scar SD99

SD DIAGNOSES AND DISEASES OF SKIN

SD01 Warts

Description
Common warts are due to an infection of the epidermis by certain human papilloma viruses, most commonly HPV subtypes 1, 2, 4, 27 and 57. They manifest typically as papillomatous, keratinous growths on the hands and feet but may affect any part of the skin (and also adjacent mucous epithelia). They are very common during childhood and adolescence.

Inclusion
plane wart
verrucae

Exclusion
genital warts GD05
molluscum contagiosum SD02

SD02 Molluscum contagiosum

Description
A disease of the skin and mucous membranes, caused by an infection with molluscum contagiosum virus. This disease is characterised by papular skin eruptions, commonly 2–3 millimetres in diameter. Transmission is by direct contact.

Exclusion
warts SD01

SD03 Herpes zoster

Description
Grouped vesicular eruptions, unilateral distribution, normally over area of a single dermatome caused by the reactivation of a latent infection with varicella zoster virus. This disease commonly presents with a rash, cutaneous hyperaesthesia or fever.

Inclusion
herpes zoster SD03.00
post-herpetic neuralgia SD03.01
shingles

Coding hint
For coding the problem level, consider Pain functions 2F84.

rash localised SS05
skin pain SS01

SD04 Herpes simplex

Description
Vesicles with erythematous base in localised area(s); plus past history of similar lesions, or virological or serological evidence.

Inclusion
cold sore
fever blister
herpes (simplex) labialis SD04.00

Exclusion
genital herpes GD03
herpes simplex of eye without corneal ulcer FD03

Coding hint
rash localised SS05

SD05 Infected finger or toe

Inclusion
paronychia SD05.00
pulp space infection of finger/thumb SD05.01
pulp space infection of toe SD05.02
whitlow

Exclusion
dermatophytosis SD08
monilia/candida SD10
post-traumatic infection finger/toe SD07

SD06 Boil or carbuncle

Description
Single or multiple focal infections of skin and soft tissues most commonly centred on the hair follicle and most commonly due to **Staphylococcus aureus**.

Inclusion
abscess
boil abscess nose
furuncle SD06.00
furunculosis SD06.01

Exclusion
boil external auditory meatus HD01
boil female external genitalia GD69
boil male external genitalia GD99
erysipelas SD16

folliculitis SD99
folliculitis barbae SD08
hydradenitis SD73
lymphadenitis BD01
perianal boil DD07
pilonidal abscess SD67
superficial pustular folliculitis SD15

SD07 Post-traumatic skin infection

Inclusion
infected post-traumatic bite
infected post-traumatic wound

Exclusion
erysipelas, pyoderma SD16
impetigo SD15
surgical wound infection AD42

SD08 Dermatophytosis

Description
Pruritic scaly lesions with central clearing and small vesicles at border; or demonstration of fungus.

Inclusion
fungal skin infection
ringworm
tinea
tinea pedis SD08.00

Exclusion
bacterial folliculitis SD06
folliculitis SD99
pityriais versicolor SD09
moniliasis/candidiasis SD11
onychomycosis SD10
superficial pustular folliculitis SD15

SD09 Pityriasis versicolor

Description
A disease of the skin, caused by an infection with the fungi **Malassezia**. This disease is characterised by white, pink, fawn, brown or often coalescing lesions that may be covered with thin furfuraceous scales. This disease commonly presents on the trunk, shoulders and arms, or neck and face. Transmission is by opportunistic transmission. Confirmation is by identification of **Malassezia** in a skin sample.

SD10 Onychomycosis

Description
Fungal infection of fingernails and/or toenails due most commonly to dermatophytes tinea unguium or yeast.

Exclusion
moniliais/candidiasis skin SD11

SD11 Candidiasis skin

Description
Candidiasis is an infection caused by yeasts of the genus **Candida**. Superficial infections of the mucous membranes and skin are common.

Inclusion
candida intertrigo SD11.00
candidiasis of nails SD11.01
diaper candidiasis SD11.02
monilial intertrigo
thrush involving nails
thrush perianal region
thrush skin

Exclusion
oral thrush DD66
genital candidiasis GD08
onychomycosis SD10

SD12 Pityriasis rosea

Description
Oval, scaly eruptions along skin tension lines of trunk, with a history of a solitary lesion preceding presenting rash.

Coding hint
rash generalised SS06
rash localised SS05

SD13 Scabies and other acariasis

Description
Scabies: Intense pruritic skin lesions plus arrays of burrows on sides of palms, fingers, penis or skin folds; or demonstration of parasites or ova in lesions. A highly contagious infestation of the skin by the mite **Sarcoptes scabiei** var. **hominis**. It may result in epidemics when introduced into institutions such as schools and nursing homes. The mites burrow into the skin, favouring the extremities, genitalia and, in infants, the axillae. The characteristic widespread intensely pruritic papulovesicular rash results

largely from the host response rather than directly to burrowing by mites. Where such a response is absent as in immunosuppressed or debilitated patients, unchecked proliferation of mites results in crusted scabies. Sarcoptic mites from other mammals such as dogs may cause a transient pruritic eruption.

Coding hint
pruritus SS02

SD14 Pediculosis and other skin infestation

Description
Pediculosis refers to parasitic skin diseases caused by animals such as arthropods (i.e. mites, ticks and lice) and worms, but excluding (except) conditions caused by protozoa, fungi, bacteria and viruses, which are called infections.

Six epidermal parasitic skin diseases (EPSD) are of particular importance: scabies, pediculosis (head lice, body lice and pubic lice infestation), tungiasis (sand flea disease) and hookworm-related cutaneous larva migrans (HrCLM). They are either prevalent in resource-poor settings or are associated with important morbidity.

Inclusion
fleas
head lice SD14.00
mites
pediculosis pubis SD14.01

Exclusion
cutaneous larva migrans DD07
infected insect bites SD07
insect bites SD39

Coding hint
pruritus SS02
localised rash SS05

SD15 Impetigo

Description
Spreading skin lesion consisting of macules, vesicles, pustules or crust with underlying raw area.

Inclusion
impetigo secondary to other dermatosis

Exclusion
bacterial folliculitis SD16
folliculitis SD99
folliculitis barbae SD08

Coding hint
post-traumatic skin infection SD07

SD16 Other specified or unknown skin infection

Inclusion
acute bacterial lymphangitis
bacterial folliculitis SD16.05
cellulitis SD16.00
erysipelas SD16.01
erythrasma SD16.02
granuloma pyogenic SD16.03
granuloma teleangiectaticum SD16.04
pyoderma
strep skin infection
unknown skin infection

Exclusion
boil/carbuncle SD06
Buruli ulcer AD24
post-traumatic skin infection SD07
impetigo SD15
molluscum contagiosum SD02
acne SD76

SD25 Malignant neoplasm of skin

Description
Characteristic histological appearance.

Inclusion
basal cell carcinoma of skin SD25.00
Kaposi's sarcoma of skin SD25.01
malignant melanoma SD25.02
rodent ulcer
squamous cell carcinoma of skin SD25.03

Exclusion
premalignant lesion of skin SD29

Coding hint
neoplasm of skin unspecified as benign or malignant when histology is not available
    SD29
other malignant neoplasm (when primary site is uncertain) AD25

SD26 Lipoma

Description
A benign tumour composed of adipose (fatty) tissue.

SD27 Benign melanocytic naevus

Description
A naevus containing melanin.

Exclusion
congenital skin anomaly SD55
spider naevus KS99
strawberry naevus SD28

SD28 Haemangioma or lymphangioma

Description
Vascular or lymphatic tumour, elevated above skin and emptying on pressure. Neoplastic hemangioma are benign localised vascular neoplasm usually occurring in infancy and childhood. It is characterised by the formation of capillary-sized or cavernous vascular channels. The majority of cases are congenital.

Inclusion
angiomatous birthmark

Exclusion
congenital skin anomaly SD55

Coding hint
swelling localised SS03

SD29 Benign, uncertain or carcinoma in situ neoplasms of skin

Inclusion
benign skin neoplasm
dermatofibroma SD29.00
dermoid cyst
dysplastic naevus SD29.01
keratoacanthoma SD29.02
premalignant lesion
skin neoplasm not specified as benign or malignant when histology is not available
skin tags

Exclusion
haemangioma SD28
keloid, hyperkeratosis SD99
mole or pigmented nevus SD27

residual haemorrhoidal skin tag DD84
seborrhoeic/senile warts SD68
solar keratosis SD66

SD35 Bruise or contusion

Description
Superficial bruise/contusion with intact skin surface.

Inclusion
ecchymosis
haematoma
subungual haematoma SD35.00

Exclusion
bruise with broken skin SD36

SD36 Abrasion, scratch, blister

Description
An abrasion is a partial thickness wound caused by damage to the skin and can be superficial, involving only the epidermis, to deep, involving the deep dermis.

Inclusion
bruise if skin broken
graze

SD37 Laceration or cut

Description
A cut is typically thought of like a wound caused by a sharp object (such as a knife or a shard of glass). The term laceration implies a torn or jagged wound.

Inclusion
laceration
cut of skin/subcutaneous tissues

Exclusion
bite SD40
bruise with broken skin SD36

SD38 Other specified or unknown skin injury

Inclusion
avulsion of fingernail SD38.00
avulsion of toenail SD38.01
needle stick

puncture
unknown skin injury

Exclusion
animal or human bite SD40

SD39 Insect bite or sting

Description
When an insect bites, it releases saliva that can cause the skin around the bite to become red, swollen and itchy. The venom from a sting often also causes a swollen, itchy, red mark (a weal) to form on the skin.

Inclusion
non-toxic spider bite
tick bite SD39.00

Exclusion
bee sting AD44
infected bite SD07
pediculosis SD14
scabies SD13
toxic effects non-medical substance AD44
wasp sting AD44

SD40 Animal or human bite

Inclusion
non-toxic snake bite

Exclusion
toxic effects non-medical substance AD44
infected bite or sting SD39

SD41 Burn or scald

Inclusion
burn of all degrees
external chemical burn
scald of all degrees

Exclusion
sunburn SD66

SD42 Foreign body in skin

Inclusion
foreign body under nail

SD55 Congenital anomaly of skin

Inclusion
birthmark
ichthyosis
port wine stain of skin SD55.00
strawberry nevus of skin SD55.01

Exclusion
haemangioma/lymphangioma SD28

SD65 Corn or callosity

Description
Callosities are areas of focal hyperkeratosis due to repeated friction and pressure.
A corn is a sharply demarcated callosity occurring over a bony prominence, usually
on the foot, and is painful.

Inclusion
clavus

Exclusion
hyperkeratosis SD99
solar hyperkeratosis SD66

SD66 Solar keratosis or sunburn

Inclusion
actinic keratosis SD66.00
allergy to sunlight SD66.01
idiopathic photodermatosis SD66.02
photodermatitis SD66.03
photosensitivity
polymorphous light eruption
senile keratosis
solar hyperkeratosis

Exclusion
senile warts/seborrhoeic keratosis SD80
skin problems due to radiation or medical treatment AD42

SD67 Pilonidal cyst or fistula or both

Description
Pilonidal disease describes a spectrum of clinical presentations, ranging from asymptomatic hair-containing cysts and sinuses to large symptomatic abscesses of the sacrococcygeal area which tend to recur. It is found predominantly in white males in their second and third decades and is thought to result from penetration of hair into the tissues with the formation of sinuses and a foreign-body granulomatous response. Risk factors for pilonidal disease include male gender, Caucasian ethnicity, sitting occupations, obesity, a deep natal cleft and presence of hair within the natal cleft.

Inclusion
pilonidal abscess

Exclusion
dermoid cyst SD29

SD68 Seborrhoeic dermatitis

Description
Greasy, scaly lesions with underlying erythema on one or more areas of scalp, face, sternum, interscapular areas, around umbilicus and in body folds, not attributable to other skin disease.

Inclusion
cradle cap SD68.00
dandruff SD68.01

Exclusion
seborrhoeic keratosis/warts SD80

Coding hint
rash generalised SS06
rash localised SS05

SD69 Atopic eczema, dermatitis

Description
Pruritic exudative lesions with/without lichenification over face and neck, wrists and hands, chest, back of knees and front of elbow.

Inclusion
flexural dermatitis
infantile eczema

Exclusion
allergic dermatitis SD70
dermatitis/atopic eczema affecting external auditory meatus HD01
diaper rash SD71

Coding hint
infected atopic eczema SD15

SD70 Contact or allergic dermatitis

Description
Pruritic erythematous lesions related to exposure to chemical substance, friction and unknown causes.

Inclusion
allergic dermatitis
chemical dermatitis
contact dermatitis SD70.00
dermatitis NOS
eczema NOS
ingestion dermatitis due to drugs SD70.01
intertrigo
plant sting
skin allergy

Exclusion
allergy/allergic reaction unspecified AD46
atopic eczema SD69
contact and other dermatitis of eyelid FD02
contact/other dermatitis of external auditory meatus HD01
dermatitis artefacta/neurodermatitis SD99
diaper rash SD71
urticaria SD78

Coding hint
rash generalised SS06
rash localised SS05
pruritus SS02

SD71 Diaper rash

Description
Dermatitis, primarily of the diaper area and sparing creases.

SD72 Psoriasis

Description
Plaques with silvery scales on knees, elbows, or scalp and/or stippled/pitted nails. Psoriasis is a common, chronic, relapsing, inflammatory skin disorder characterised by abnormal epidermal keratinisation and hyperproliferation. It has a strong genetic component and affects some 2% of the populations of many regions of the world. Up to 10–20% of patients with psoriasis also experience an inflammatory polyarthritis (psoriatic arthritis).

Note
Double code psoriatic arthritis LD99.

SD73 Sweat gland disease

Inclusion
anhidrosis SD73.00
dyshidrosis
dyshidrotic eczema SD73.01
heat rash
hydradenitis SD73.02
miliaria
pompholyx
prickly heat
sweat rash

Exclusion
hyperhidrosis AS10

SD74 Sebaceous cyst

Description
Intradermal or subcutaneous sac-like structure, the wall of which is stratified epithelium containing keratohyalin granules.

Inclusion
atheroma cyst SD74.00
epidermoid cyst
epithelial cyst SD74.01
pilar cyst
trichilemmal cyst

Exclusion
other cutaneous cyst SD99

SD75 Ingrowing nail

Inclusion
ingrowing nail with infection

Exclusion
paronychia SD05

SD76 Acne

Description
A group of related disorders characterised by follicular occlusion and inflammation.

Inclusion
acne conglobata SD76.00
acne vulgaris SD76.01
blackheads
comedones
pimples

Exclusion
acne due to medication AD41

SD77 Chronic ulcer of skin

Description
A skin ulcer is an open wound that develops on the skin as a result of injury, poor circulation or pressure. Skin ulcers can take a very long time to heal.

Inclusion
bedsore
diabetic foot ulcer SD77.02
decubitus ulcer
pressure sore SD77.00
varicose ulcer
venous ulcer of leg SD77.01

Exclusion
gangrene KD67

SD78 Urticaria

Description
A vascular reaction of the skin characterised by erythema and wheal formation due to localised increase of vascular permeability. The causative mechanism may be allergy, infection or stress.

Inclusion
hives
weals

Exclusion
angioedema/allergic oedema AD46
drug allergy AD41

SD80 Seborrhoeic keratosis

Description
Seborrhoeic keratoses are very common benign neoplasms of epidermal keratinocytes which increase in prevalence and number with age. They are commonly multiple and are very variable in shape and colour.

SD81 Rosacea

Description
Rosacea encompasses a spectrum of changes that occur mainly in facial skin but may also involve the eyes. Most patients with rosacea have facial erythema and vascular instability which are variably associated with inflammatory papules and pustules, hypertrophic changes and ocular involvement.

Inclusion
perioral dermatitis
rhinophyma

SD82 Alopecia

Inclusion
alopecia areata SD82.00
androgenic alopecia SD82.01

SD99 Other specified or unknown diagnoses and diseases of skin

Inclusion
dermatitis artefacta
discoid lupus erythematosus SD99.00
erythema multiforme
erythema nodosum SD99.01
folliculitis
granulomatosis
granuloma annulare
hyperkeratosis NOS
keloid SD99.02
lichen planus SD99.03
lichen sclerosus SD99.04

neurodermatitis
onychogryphosis SD99.05
pemphigus
pigmentation
scar
striae atrophicae SD99.06
vitiligo SD99.07

Exclusion
bacterial folliculitis SD06
folliculitis barbae SD08
superficial pustular folliculitis SD15

T ENDOCRINE, METABOLIC AND NUTRITIONAL SYSTEM

TS SYMPTOMS, COMPLAINTS AND ABNORMAL FINDINGS OF ENDOCRINE, METABOLIC AND NUTRITIONAL SYSTEM

TS01 Excessive thirst

Description
A thirst that a person cannot quench by drinking.

Inclusion
polydipsia

Coding hint
For coding the problem level, consider Drinking 2F39.

TS02 Excessive appetite

Description
Intermittent or persistent increased drive (urge) or desire to eat food as compared to what is typical for the individual.

Inclusion
overeating
polyphagia

Exclusion
bulimia PD17

Coding hint
For coding the problem level, consider Eating 2F38.

TS03 Loss of appetite

Description
Intermittent or persistent decreased motivation or desire to eat food as compared to what is typical for the individual. Anorexia is a pathological lack or loss of appetite.

Inclusion
anorexia

Exclusion
anorexia nervosa PD17
cachexia TS07

Coding hint
For coding the problem level, consider Eating 2F38.

TS04 Feeding problem of infant or child

Inclusion
problem of how to feed infant or child

Exclusion
breast feeding problems WS06
feeding problem/eating disorders with psychological cause PS09
food allergy AD46
food intolerance DD99

TS05 Feeding problem of adult

Inclusion
problem of what and how to eat/feed adult

Exclusion
anorexia/bulimia nervosa PD17
dysphagia DS21
food allergy AD46
food intolerance DD99
loss of appetite TS03
psychological eating disorders/ food refusal PS99

Coding hint
For coding the problem level, consider Eating 2F38.

TS06 Weight gain

Description
An increase in body weight.

Exclusion
obesity TD66
overweight TS51

TS07 Weight loss

Description
A decrease in body weight.

Inclusion
cachexia

Exclusion
anorexia nervosa PD17

TS08 Growth delay

Description
Delay of expected physiological development includes delayed milestone of development as normal within the appropriate cultural environment including gross and fine motor development, language, social/cultural milestones.

Inclusion
failure to thrive
physiological delay growth

Exclusion
delayed milestones PS18
delayed puberty TD99
learning disorder PS20
mental retardation PD18

TS09 Dehydration

Description
Dehydration occurs when there is an insufficient amount or excessive loss of water in the body. This can be caused by vomiting, diarrhoea, fever, use of diuretics, profuse sweating or decreased water intake.

Inclusion
water depletion

Exclusion
salt depletion/electrolyte disturbance TD99

TS50 Underweight

Description
A weight below a weight considered normal or desirable.

TS51 Overweight

Description
Overweight is a condition characterised by excess weight relative to height. Overweight is assessed by the body mass index (BMI). The BMI is a measure of body mass relative to height, calculated as weight $(kg)/height^2$ (m^2). The BMI categories for defining overweight vary by age and gender in infants, children and adolescents. For adults, overweight is defined by a BMI ranging from 25.00 to 29.99 kg/m^2.

Exclusion
obesity TD66

TS90 Concern or fear of disease of endocrine, metabolic and nutritional system

Description
Concern about/fear of other endocrine, metabolic or nutritional disease in a patient without the disease, until the diagnosis is proven.

Inclusion
concern of diabetes
concern of cancer of the endocrine, metabolic system
fear of diabetes
fear of cancer of the endocrine, metabolic system

Coding hint
If the patient has the disease, code the disease.

TS99 Other specified endocrine, metabolic, nutritional symptoms, complaints, abnormal findings

Inclusion
specific food craving

Exclusion
hyperglycemia AD23
fluid retention KS04

TD DIAGNOSES AND DISEASES OF ENDOCRINE, METABOLIC AND NUTRITIONAL SYSTEM

TD01 Endocrine infection

Exclusion
auto-immune thyroiditis TD99
drug-induced thyroiditis TD99
subacute thyroiditis TD99

TD25 Malignant neoplasm of thyroid

Description
Characteristic histological appearance.

TD26 Benign neoplasm of thyroid

Exclusion
endocrine neoplasm, other specified TD27
goitre TD65

TD27 Other specified or unknown endocrine neoplasm

Inclusion
in situ endocrine neoplasm of endocrine system
neoplasm of unknown or uncertain behaviour of endocrine system TD27.00
other benign endocrine neoplasm of endocrine system TD27.01
other malignant endocrine neoplasm of endocrine system TD27.02

Exclusion
malignant neoplasm of thyroid TD25
benine neoplasm of thyroid TD26

TD55 Thyroglossal duct or cyst

Description
A cyst or duct in the neck caused by persistence of portions of, or by lack of closure of, the primitive thyroglossal duct.

Exclusion
goitre TD65

TD56 Congenital anomaly of endocrine or metabolic system

Inclusion
cretinism
dwarfism

Exclusion
thyroglossal duct (cyst) TD55

TD65 Goitre

Description
Enlargement of the thyroid gland due to follicular multiplication, unaccompanied by hyperthyroidism or thyrotoxicosis.

Inclusion
non-toxic goitre
thyroid nodule

Exclusion
benign neoplasm thyroid TD26
hypothyroidism TD69
malignant neoplasm thyroid TD25
neoplasm endocrine other/unspecified TD27
thyroglossal cyst TD55
toxic goitre TD68

TD66 Obesity

Description
Obesity is defined as a body mass index (BMI) greater than or equal to 30.00 kg/m². There are three levels of severity in recognition of different management options.

Exclusion
overweight TS51

Note
There are 3 classes BMI 30.00–34.9 obesity class 1 (low risk), obesity class 2 (moderate risk) 35.00–39.99 and third class (high risk) greater or equal 40.00 = morbid obesity.

TD68 Hyperthyroidism or thyrotoxicosis

Description
A hypermetabolic condition associated with elevated levels of free thyroxine and/or free triiodothyronine resulting in excess synthesis and secretion of thyroid hormone.

Inclusion
Graves' disease
toxic goitre

Exclusion
non-toxic goitre TD65
Hashimoto's thyrotoxicosis TD99

Coding hint
For coding the problem level, consider Energy level 2F71.

TD69 Hypothyroidism or myxoedema

Description
Laboratory evidence of diminished thyroid hormone activity and excessive thyroid stimulating hormone; or four or more of the following: weakness/tiredness; mental changes: apathy, poor memory, slowing; voice changes: coarser, deeper slower speech; undue sensitivity to cold; constipation; coarse puffy facial features; cool, dry, sallow skin, decreased sweating; peripheral oedema.

Exclusion
cretinism TD56

Coding hint
other complaint of metabolism TS99
For coding the problem level, consider Energy level 2F71.

TD70 Hypoglycaemia

Description
Hypoglycaemia demonstrated by biochemical testing, or characteristic symptoms in a diabetic patient relieved by ingestion or injection of sugar.

Inclusion
hyperinsulinism
insulin coma

TD71 Type 1 diabetes mellitus

Description
Diabetes mellitus type 1 (type 1 diabetes, T1DM, formerly insulin-dependent or juvenile diabetes) is a form of diabetes mellitus that results from destruction of insulin-producing beta cells, mostly by autoimmune mechanisms. The subsequent lack of insulin leads to increased blood and urine glucose.

Exclusion
drug-induced hyperglycaemia AD41
hyperglycaemia as isolated finding AD23
type 2 diabetes TD72
gestational diabetes WD72

Note
1. Double code complications such as retinopathy FD67, nephropathy UD65.
2. In pregnancy, double code with WD71.

TD72 Type 2 diabetes mellitus

Description
Diabetes mellitus type 2 (formerly non-insulin-dependent diabetes mellitus (NIDDM) or adult-onset diabetes) is a metabolic disorder that is characterised by high blood glucose in the context of insulin resistance and relative insulin deficiency.

Inclusion
diabetes NOS

Exclusion
drug-induced hyperglycaemia AD41
hyperglycaemia as isolated finding AD23
type 2 diabetes TD71
gestational diabetes WD72

Note
1. Double code complications such as retinopathy FD67, nephropathy UD65.
2. In pregnancy, double code with WD71.

TD73 Vitamin deficiency

Inclusion
beriberi/vitamin B1 deficiency
scurvy
vitamin D deficiency with rickets
vitamin D deficiency without rickets

Exclusion
anaemia vit B12/folate deficiency BD67

TD74 Mineral and nutritional deficiency

Inclusion
dietary mineral deficiency
iron deficiency without anaemia
kwashiorkor TD74.00
marasmus TD74.01
malnutrition

Exclusion
iron deficiency anaemia BD66
malabsorption syndrome/sprue DD99
pernicious anaemia BD67

TD75 Lipid disorder

Inclusion
abnormality of lipoprotein level
hypercholesterolaemia TD75.00
hypertriglyceridaemia TD75.01
mixed hyperlipidaemia TD75.02
primary hypercholesterolaemia TD75.03
raised level of cholesterol/triglycerides
xanthoma

TD99 Other specified or unknown endocrine, metabolic, nutritional diagnoses and diseases

Inclusion
acromegaly
Addison's disease TD99.00
adrenal/ovarian/pituitary/parathyroid/testicular/other endocrine dysfunction
adrenocortical insufficiency TD99.01
adrenogenital disorder TD99.02
amyloidosis
Cushing's syndrome TD99.03
diabetes insipidus
Gilbert's syndrome
hyperaldosteronism
hyperhomocysteinemia TD99.04
lactose intolerance TD99.05
polycystic ovary syndrome TD99.06
porphyria TD99.07
precocious/delayed puberty
premature menopause TD99.08
pubertas praecox TD99.09
raised uric acid
renal glycosuria
thyroiditis TD99.10

Exclusion
food allergy AD46
food intolerance DD99
infectious thyroiditis TD01
osteoporosis LD81

U URINARY SYSTEM

US SYMPTOMS, COMPLAINTS AND ABNORMAL FINDINGS OF URINARY SYSTEM

US01 Dysuria or painful urination or both

Description
Dysuria is characterised by painful urination.

Inclusion
burning urination
strangury
vesical tenesmus

Exclusion
frequent/urgent urination US02
urethritis UD03

US02 Urinary frequency or urgency

Description
Urinary frequency is the need to urinate many times during the day, at night (nocturia) or both but in normal or less-than-normal volumes. Frequency may be accompanied by a sensation of an urgent need to void (urinary urgency).

Polyuria: Polyuria has generally been defined as a urine output exceeding 3 L/day in adults and 2 L/m^2 in children. It must be differentiated from the more common complaints of frequency or nocturia, which may not be associated with an increase in the total urine output.

Inclusion
nocturia
polyuria

US03 Urine incontinence

Description
Any condition of the urinary system, caused by determinants arising during the antenatal period or after birth, leading to loss of voluntary control or support of the urethra. These conditions are characterised by involuntary leakage of large amounts of urine in association with uninhibited contractions of the detrusor muscle and the inability to control urination.

Inclusion
enuresis of organic origin
genuine stress incontinence US03.00
involuntary urination
mixed incontinence US03.01
stress incontinence

urge incontinence US03.02

Exclusion
urine incontinence of psychogenic origin PS10

US04 Urinary retention

Description
Incomplete emptying of the bladder.

US05 Other specified urination problems

Inclusion
anuria US05.00
dribbling urine
oliguria US05.00

Exclusion
urinary retention US04

US06 Haematuria

Description
Haematuria is characterised by the presence of red blood cells (RBCs) in the urine.

Inclusion
blood in urin
microscopic haematuria

Exclusion
abnormal urine test US50

US07 Other specified urine symptom or complaint

Inclusion
dark urine
malodorous urine

Exclusion
abnormal urine test US50

US08 Other specified bladder symptom or complaint

Inclusion
bladder pain
irritable bladder

US09 Kidney symptom or complaint

Inclusion
kidney pain
kidney trouble
renal colic US09.00

Exclusion
loin/flank pain LS05

US10 Urethral discharge

Description
Urethral discharge is any type of discharge or liquid, besides urine or semen, that comes out of the opening of the urethra.

US50 Abnormal urine test

Inclusion
asymptomatic bacteriuria
glycosuria US50.01
orthostatic albuminuria US50.00
proteinuria US50.02
pus in urine
pyuria

Exclusion
haematuria/blood in urine US06

US90 Concern or fear of disease of urinary system

Description
Concern about/fear of other urinary disease in a patient without the disease, until the diagnosis is proven.

Coding hint
If the patient has the disease, code the disease.

US99 Other specified symptom, complaint and abnormal finding of urinary system

Exclusion
irritable bladder/bladder pain US08
kidney symptom/complaint US09

UD DIAGNOSES AND DISEASES OF URINARY SYSTEM

UD01 Pyelonephritis or pyelitis

Description
Acute pyelonephritis is characterised by an inflammation of the renal pelvis and paren-
chyma due to bacterial infection. Symptoms include fever, loin (kidney) pain, nausea
and vomiting. Concurrently, symptoms of acute cystitis with dysuria, frequency and
haematuria may occur.

Inclusion
infection of kidney
renal or perinephric abscess
tubulo-interstitial nephritis

Coding hint
cystitis/other urinary infection UD02

UD02 Cystitis

Description
A condition of the bladder caused by infection, reaction to pharmacological agents,
exposure to radiation therapy or potential irritants. This condition is characterised by
inflammation of the urinary bladder, dysuria, pollakiuria, fever or flank pain.

Inclusion
acute cystitis (non-veneral) UD02.00
chronic cystitis (non-veneral)
interstitial (chronic) cystitis

Exclusion
balanitis GD08
prostatitis GD10
pyelonephritis UD01
urethritis UD03
vaginitis GD12

Coding hint
Consider US01 and US02.
In pregnancy, also code WD71.

UD03 Urethritis and urethral syndrome

Description
A condition characterised by inflammation or irritation of the urethra.

Inclusion
meatitis
non-specific urethritis

Exclusion
gonococcal urethritis female GD02
gonococcal urethritis male GD02
Reiter disease; urethrotrigonitis UD02
urethritis trichomonal female GD04
urethritis chlamydial female GD06

Coding hint
frequent/urgent urination US02
irritable bladder US08
painful urination US0
urethral discharge US10

UD04 Other specified or unknown urinary infection

Description
An infection of the kidney, ureter or urethra caused by microbes.

Inclusion
lower urinary tract infection
urinary tract infection NOS UD04.00

UD25 Malignant neoplasm of kidney

Description
Characteristic histological appearance.

Coding hint
uncertain or carcinoma in situ neoplasm of urinary tract UD29

UD26 Malignant neoplasm of bladder

Description
Characteristic histological appearance.

Coding hint
uncertain or carcinoma in situ neoplasm of urinary tract UD29

UD27 Other specified and unknown malignant neoplasm urinary tract

Description
Characteristic histological appearance.

Inclusion
malignant neoplasm ureter
malignant neoplasm urethra
unknown malignant neoplasm urinary tract

Exclusion
malignant neoplasm prostate GD26

Coding hint
uncertain or carcinoma in situ neoplasm of urinary tract UD29

UD28 Benign neoplasm of urinary tract

Description
Characteristic histological appearance.

Inclusion
polyp of urinary tract
polyp of urine bladder UD28.00

Exclusion
prostatic hypertrophy GD70

Coding hint
uncertain or carcinoma in situ neoplasm of urinary tract UD29

UD29 Uncertain or carcinoma in situ neoplasm of urinary system

UD35 Injury to urinary tract

Inclusion
contusion of kidney UD35.00
foreign body in urinary tract UD35.01

UD55 Congenital anomaly of urinary system

Inclusion
congenital polycystic kidney disease UD55.00
congenital single renal cyst
congenital urethral valves
duplex kidney/ureter

UD65 Nephrosis

Description
A non-inflammatory disease of the kidneys chiefly affecting function of the nephrons.

Inclusion
analgesic nephropathy
glomerulonephritis
nephritis
nephropathy
nephrosclerosis

nephrotic syndrome

Exclusion
renal failure UD99

Coding hint
Consider abnormal urine test US50.

Note
Double code diabetic nephropathy with TD71 and TD72.

UD66 Chronic kidney disease

Description
Glomerular filtration rate (GFR) less than 60 or presence of kidney damage that is present for more than 3 months.

Inclusion
chronic renal failure
chronic renal insufficiency UD66.00

Exclusion
acute kidney failure UD99

Coding hint
For coding the problem level, consider Energy level 2F71.

Note
Double code known causative disease.

UD67 Urinary calculus

Description
Urinary calculus is characterised by colicky pain and either haematuria or history of urinary stone in the past, or passage of calculus or imaging evidence of calculus.

Inclusion
stone in bladder
stone in kidney
stone in ureter
urolithiasis

Coding hint
abnormal urine test US50
blood in urine US06
other urinary symptom US99
renal colic US09

UD99 Other specified or unknown diagnoses and diseases of urinary tract

Inclusion
acute renal failure
bladder diverticulum
contracted kidney UD99.00
hydronephrosis
hypertrophic kidney
obstruction in bladder neck
obstructive vesicoureteric reflux UD99.01
ureteric reflux
urethral caruncle
urethral stricture UD99.02

W PREGNANCY AND CHILDBEARING

WS SYMPTOMS, COMPLAINTS AND ABNORMAL FINDINGS DURING PREGNANCY, DELIVERY AND PUERPERIUM
WS01 Suspicion of pregnancy

Inclusion
delayed menstruation
symptoms suggestive of pregnancy

Exclusion
fear of pregnancy WS90
pregnancy confirmed WD67
unwanted pregnancy WD68

WS02 Pregnancy vomiting and nausea

Inclusion
hyperemesis gravidarum WS02.00
morning sickness in confirmed pregnancy

WS03 Bleeding first 20 weeks of pregnancy

Description
Bleeding during pregnancy before 21 weeks of pregnancy.

Inclusion
bleeding first trimester WS03.00
implantation bleeding, a minimal haemorrhage seen at the time of implantation of
   the egg

Exclusion
antepartum haemorrhage WS04

Coding hint
spontaneous abortion WD65

WS04 Antepartum haemorrhage

Description
Bleeding from the uterus during a pregnancy after the 20th week.

Inclusion
bleeding second/third trimester WS04.00

WS05 Post-partum bleeding

Description
Heavy bleeding at or within 6 weeks of parturition.

WS06 Breast or lactation symptom or complaint

Inclusion
galactorrhoea associated with childbirth
lactation problem WS06.00
suppressed lactation
suppression of lactation
weaning

Exclusion
cracked nipples WD84
puerperal mastitis WD03

WS39 Other specified post-partum symptom or complaint

Description
Complaints related to and within 6 weeks of parturition.

Inclusion
abnormal lochia WS39.00

Exclusion
lactation complaints WS06
complications of puerperium WD85
puerperal depression PD12
post-partum bleeding WS05

WS50 Abnormal findings on antenatal screening of mother

WS90 Concern or fear of being pregnant

Description
Concerns about or fear of being pregnant without the pregnancy being proven.

Inclusion
concern about possibility of unwanted pregnancy

Exclusion
concern/fear if unwanted pregnancy confirmed WD68

WS91 Fear about complications of pregnancy

Description
Concern about/fear of complications in a patient without them, until they are proven.

Inclusion
fear of congenital anomaly in baby

Coding hint
If the patient has the complication, code the complication.

WS99 Other specified symptoms, complaints and abnormal findings during pregnancy, delivery and puerperium

Inclusion
concern about appearance during pregnancy
feeling fewer movements of fetus WS99.00
pelvic instability WS99.01

WD DIAGNOSES DURING PREGNANCY, DELIVERY AND PUERPERIUM

WD01 Puerperal infection or sepsis

Description
Infection of birth canal or reproductive organs within 6 weeks of parturition.

Inclusion
puerpural endometritis WD01.01
infection of caesarean section wound
infection of perineal wound WD01.00

Exclusion
obstetric tetanus ND03

WD02 Other specified and unknown infection complicating pregnancy, delivery and puerperium

Inclusion
genitourinary tract infection in pregnancy WD02.00

Exclusion
puerperal infection WD01
puerperal mastitis WD03

WD03 Puerperal mastitis

Description
Pain, inflammation of breast within 6 weeks of parturition or while lactating.

Inclusion
breast abscess

Exclusion
inflammatory disorders breast GD99

Coding hint
disorders of lactation WS06

WD25 Malignant neoplasms related to pregnancy

Description
Characteristic histological appearance.

Inclusion
choriocarcinoma
chorioepithelioma

WD26 Benign, in situ or uncertain neoplasms related to pregnancy

Description
Benign, in situ or uncertain neoplasm related to pregnancy; hydatidiform mole; neoplasm related to pregnancy not specified as benign or malignant when histology is not available

Inclusion
hydatidiform mole

WD35 Injury complicating pregnancy

Inclusion
results of injury interfering with pregnancy

Exclusion
new injury caused by childbirth (complicated labour/delivery livebirth) WD82
new injury caused by childbirth (complicated labour/delivery stillbirth) WD83

Coding hint
Consider coding the manifestation of the injury.

WD55 Congenital anomaly complicating pregnancy

Inclusion
maternal anomaly which could affect pregnancy/childbirth

Exclusion
foetal anomaly affecting pregnancy and childbirth WD71

WD65 Spontaneous abortion

Description
Miscarriage, also known as spontaneous abortion and pregnancy loss, is characterised
by non-induced embryonic or foetal death or passage of products of conception prior
to 20 weeks gestation or weighing less than 500 grams.

Inclusion
complete abortion
habitual abortion WD65.00
incomplete abortion
miscarriage
missed abortion
recurrent abortion

Exclusion
antepartum bleeding WS03 and WS04
foetal death/stillbirth after the 28th week of pregnancy WD83
induced abortion WD66

WD66 Induced abortion

Inclusion
termination of pregnancy, with or without complications

Exclusion
Abortion, spontaneous WD65

Note
Contrary with complications of a pregnancy and delivery that are coded separately,
complications of an induced abortion are included in this class.

WD67 Pregnancy

Inclusion
confirmed pregnancy

Exclusion
ectopic pregnancy WD69
high-risk pregnancy WD71
unwanted pregnancy WD68

WD68 Unwanted pregnancy

Description
Unwanted pregnancy is a pregnancy that is not desired.

WD69 Ectopic pregnancy

Description
Pregnancy in a place other than inside the uterus. Confirmation by ultrasound, laparoscopy, culdoscopy or surgery.

WD70 Pre-eclampsia or eclampsia

Inclusion
haemolysis elevated liver enzymes low platelet count syndrome WD70.00
pregnancy-induced hypertension complicating pregnancy, childbirth or the puerperium WD70.01
proteinuria and oedema in pregnancy
toxaemia/(pre) eclampsia in pregnancy WD70.02

Exclusion
pre-existing hypertension KD73

Coding hint
Pregnancy symptom/complaint, other WS99

WD71 Pregnancy, high risk

Description
A 'high-risk' pregnancy means a woman has one or more risk factors that raise her – or her baby's – chances for health problems or preterm (early) delivery.

Inclusion
abnormal foetal presentation WD71.00
aged primipara
anaemia of pregnancy
cervical insufficiency/incompetence WD71.01
foetal-maternal disproportion WD71.02

foetal growth retardation WD71.03
history of recurrent miscarriages
malpresentation
multiple gestation
multiple pregnancy
placenta praevia
polyhydramnios
pre-existing diabetes mellitus in pregnancy WD71.04
pre-existing hypertension WD71.05
premature labour
previous caesarean section
rhesus antibody present WD71.06
small foetus for age

Exclusion
infections complicating pregnancy WD02
ectopic pregnancy WD69
gestational diabetes WD72
pre-eclampsia/eclampsia WD70

WD72 Gestational diabetes

Description
Diabetes mellitus arising or diagnosed in pregnancy (per WHO criteria or other national criteria). Gestational diabetes mellitus is defined as any degree of glucose intolerance with onset or first recognition during pregnancy. The definition applies regardless of whether insulin or only diet modification is used for treatment or whether the condition persists after pregnancy.

Inclusion
diabetes manifested during pregnancy

Exclusion
pre-existing diabetes (type 1 diabetes melitus) TD71
pre-existing diabetes (type 2 diabetes melitus) TD72

Coding hint
hyperglycaemia AD23

WD80 Uncomplicated labour, delivery livebirth

Description
Definition of normal labour encompasses features such as spontaneous onset, low risk at the start and remaining so throughout the process. The neonate is born spontaneously in the vertex presentation between 37 and 42 completed weeks of pregnancy. After birth, mother and infant are in good condition.

Note
For the intervention, use code -215.

WD81 Uncomplicated labour, delivery stillbirth

Description
The delivery of a foetus that has died in the womb (strictly, after having survived through at least the first 20 weeks of pregnancy, earlier instances being regarded as abortion or miscarriage).

Note
For the intervention, use code -215.

WD82 Complicated labour, delivery livebirth

Inclusion
assisted extraction of livebirth
breech delivery livebirth
caesarean section of livebirth WD82.00
delivery by vacuum extraction of livebirth WD82.02
dystocia livebirth
forceps delivery of livebirth WD82.04
induction of labour of livebirth
injuries caused by childbirth
livebirth after complicated delivery
placenta praevia in delivery of livebirth

Exclusion
antepartum haemorrhage WS04
post-partum haemorrhage WS05
pre-eclampsia/eclampsia WD70

Note
For the intervention, use code -215.

WD83 Complicated labour, delivery stillbirth

Inclusion
assisted extraction of stillbirth
breech delivery stillbirth
caesarean section of stillbirth WD83.00
delivery by vacuum extraction of stillbirth WD83.01
dystocia stillbirth
forceps delivery of stillbirth WD83.02
induction of labour stillbirth
injuries caused by childbirth
placenta praevia in delivery stillbirth
stillbirth after complicated delivery

Exclusion
post-partum haemorrhage WS05
pre-eclampsia/eclampsia WD70

Note
For the intervention, use code -215.

WD84 Other specified and unknown breast disorder in pregnancy or puerperium

Inclusion
breast disorder in puerperium
cracked nipple WD84.00
unknown breast disorder in pregnancy or puerperium

Exclusion
breast/lactation symptom/complaint WS06
puerperal mastitis WD03

WD85 Other specified complications of puerperium

Inclusion
haemorrhoids in puerperium WD85.00
sub-involution of uterus WD85.01
thrombosis complicating pregnancy and/or puerperium WD85.02

Exclusion
puerperal depression PD12
puerperal psychosis PD06
puerperal infection WD01
pre-eclampsia/eclampsia WD69
breast disorder in pregnancy WD84
disruption of episiotomy wound in the puerperium AD42

WD99 Other specified and unknown diagnoses and diseases or health conditions in pregnancy, delivery and puerperium

Inclusion
deep venous thrombosis in pregnancy WD99.02
false labour WD99.00
haemorrhoids in pregnancy WD99.03
prolonged pregnancy WD99.01
varicose veins in pregnancy WD99.04

Exclusion
pseudocyesis PD99

Z SOCIAL PROBLEMS

Description
Classes in this chapter are provided for occasions when circumstances other than a disease, injury or external cause classifiable elsewhere are recorded as 'diagnoses' or 'problems'.

ZC SOCIAL PROBLEMS INFLUENCING HEALTH STATUS
Description
A social problem is an issue within the personal environment or society that makes it difficult for people to achieve their full potential. Poverty, unemployment, unequal opportunity, racism and malnutrition are examples of social problems. So are substandard housing, employment discrimination, and child abuse and neglect.

ZC01 Partner relationship problem

Description
Partner relationship problems related to the way in which partners feel and behave towards each other.

Inclusion
emotional abuse

Exclusion
physical abuse by partner ZC30
victim of physical abuse ZC35

Note
The diagnosis of problems in the relationship between family partners requires the patient's agreement on the existence of the problem and desire for help.

ZC02 Child relationship problem

Description
Child relationship problems are problems in the way in which a parent and his/her child feel and behave towards each other.

Inclusion
emotional child abuse
neglected child ZC02.00

Exclusion
physical abuse ZC35

Note
The diagnosis of problems in the relationship with a child requires the patient's agreement on the existence of the problem and desire for help.

ZC03 Parent or family member relationship problem

Description
Parent and family relationship problems are problems in the way in which a person and parents or other family feel and behave towards each other (Cambridge dictionary).

Inclusion
relationship problem with adult
relationship problem with parent
relationship problem with sibling
relationship problem with other family member

Exclusion
relationship problem with partner ZC01
relationship problem with child ZC02
relationship problem with friend ZC09

Note
The diagnosis of problems in the relationship between family members requires the patient's agreement on the existence of the problem and desire for help.

ZC04 Health care provider relationship problem

Description
Health care provider relationship problems are problems in the way in which a person and health care providers behave towards each other.

Inclusion
doctor/patient problems

Note
The diagnosis of problems in the relationship with a health provider requires the patient's agreement on the existence of the problem and desire for help.

ZC09 Other specified relationship problem

Description
Other relationship problems are problems in the way in which two or more people feel and behave towards each other.

Inclusion
relationship problems with friends
neighbours' quarrel/noise ZC09.00
relationship problems with neighbours

Exclusion
relationship problem with family member ZC03

Coding hint
For qualifying the level of the problem, use 2F53 in addition.

Note
The diagnosis of problems in the relationship with friends requires the patient's agreement on the existence of the problem and desire for help.

ZC10 Loss or death of partner problem

Description
Problem related to loss or death of partner.

Inclusion
bereavement
divorce from partner ZC10.00
death of partner ZC10.01

Note
The diagnosis of problems arising from the loss or death of a partner requires the patient's agreement on the existence of the problem and desire for help.

ZC11 Loss or death of child problem

Description
Problem related to loss or death of child.

Note
The diagnosis of problems arising from the loss or death of a child in the family requires the patient's agreement on the existence of the problem and desire for help.

ZC12 Loss or death of parent or family member problem

Description
Problem related to loss or death of parent or family member.

Exclusion
loss of child ZC11
loss of partner ZC10

Note
The diagnosis of problems arising from the loss or death of a family member requires the patient's agreement on the existence of the problem and desire for help.

ZC13 Problems associated with finances

Inclusion
financial problem
poverty

Note
The diagnosis of problems associated with finances requires acknowledgement of existence of the problem and desire for help.

ZC15 Education problem

Description
Problem related to education and literacy.

Inclusion
illiteracy ZC15.00
failed exams ZC15.01
low literacy
poor educational progress ZC15.02

Note
The diagnosis of problems with education essentially requires the patient's expression of concern about them, with agreement about the existence of the problem and desire for help. Whatever the objective education status, patients can consider this as a problem. Labelling these problems requires acknowledgement of absolute differences in education, as well as the individual's perception.

ZC16 Work problem

Description
Problems related to employment.

Inclusion
discord in workplace ZC16.00
occupational exposure to toxic agents ZC16.01
occupational noise exposure ZC16.02
stressful work schedule ZC16.03
threat of dismissal ZC16.04

Note
The diagnosis of problems with working conditions essentially requires the patient's expression of concern about them, with agreement about the existence of the problem and desire for help. Whatever the objective working conditions, patients can consider these as a problem. Labelling these problems requires acknowledgement of absolute differences in working conditions, as well as the individual's perception.

ZC17 Unemployment problem

Description
Problems related to unemployment

Exclusion
problems related to employment ZC16

Note
The diagnosis of problems with unemployment essentially requires the patient's expression of concern about them, with agreement about the existence of the problem and desire for help. Whatever the objective nature of the unemployment, patients can consider this as a problem. Labelling these problems requires acknowledgement of absolute differences in unemployment, as well as the individual's perception.

ZC20 Food or water problem

Description
Problems related to food and water.

Note
The diagnosis of problems with food/water conditions essentially requires the patient's expression of concern about them, with agreement about the existence of the problem and desire for help. Whatever the objective or nature of food/water conditions, patients can consider these as a problem. Labelling these problems requires acknowledgement of absolute differences in food/water conditions, as well as the individual's perception.

ZC25 Illness of partner problem

Description
Problems related to an illness in the patient's partner.

Note
The diagnosis of problems arising from a partner being ill requires the patient's agreement on the existence of the problem and desire for help.

ZC26 Illness of child problem

Description
Problems related to the illness of a child.

Note
The diagnosis of problems arising due to a child being ill requires the patient's agreement on the existence of the problem and desire for help.

ZC27 Illness of parents or family member problem

Description
Problems related to illness of parents or other family.

Exclusion
problem with partner being ill ZC25

Note
The diagnosis of problems arising from the illness of a family member requires the patient's agreement on the existence of the problem and desire for help.

ZC30 Partner's behaviour problem

Description
Problem with the way a partner conducts him/herself or behaves.

Inclusion
addiction of partner ZC30.00
aggressive behaviour of partner ZC30.01
infidelity of partner ZC30.02

Exclusion
victim of physical abuse ZC35

Note
The diagnosis of problems arising from the behaviour of a partner requires the patient's agreement on the existence of the problem and desire for help.

ZC31 Parent or family behaviour problem

Description
Problem with the way a parent/family member conducts him/herself or behaves.

Inclusion
addiction of parent or family ZC31.00
aggression of parent or family ZC31.01

Exclusion
problem with behaviour partner ZC30

Note
The diagnosis of problems arising from the behaviour of a family member requires the patient's agreement on the existence of the problem and desire for help.

ZC35 Violence problem

Description
Victim of physical abuse, violence, rape, sexual attack.

Inclusion
maltreatment/sexual abuse of child ZC35.00
maltreatment/sexual abuse by partner ZC35.01
problems related to assault/rape ZC35.02

victim of physical abuse
victim of rape
victim of sexual attack

Exclusion
child emotional abuse ZC02
partner emotional abuse ZC01
partner physical abuse ZC30
physical problems to be coded in appropriate rubric(s) in other chapters
psychological problems to be coded in Chapter P

Note
The diagnosis of social problems arising from assaults and other harmful events requires the patient's agreement on the existence of the problem and desire for help.

ZC36 Housing problem

Description
Problems related to housing conditions.

Inclusion
accommodation unsuitable
homeless ZC36.00
housing unsuited to needs ZC36.01
inadequate housing

Note
The diagnosis of problems with housing conditions essentially requires the patient's expression of concern about them, with agreement about the existence of the problem and desire for help. Whatever the objective housing conditions, patients can consider these as a problem. Labelling these problems requires acknowledgement of absolute differences in housing conditions, as well as the individual's perception.

ZC37 Legal problem

Description
Problems concerning the legislation and other law of a country.

Inclusion
arrest
incarceration
imprisonment ZC37.00
prosecution
problems related to release from prison
problems with guardianship ZC37.01

Note

The diagnosis of problems with legal issues essentially requires the patient's expression of concern about them, with agreement about the existence of the problem and desire for help. Whatever the objective legal issues, patients can consider these as a problem. Labelling these problems requires acknowledgement of absolute differences in legal issues as well as the individual's perception.

ZC38 Social welfare problem

Description

Problems related to social insurance and (the lack of) welfare care by the government.

Inclusion

sickness and disability law problem ZC38.00
social assistance law problem ZC38.01

Note

The diagnosis of problems with social welfare essentially requires the patient's expression of concern about them, with agreement about the existence of the problem and desire for help. Whatever the objective social welfare situation, patients can consider this as a problem. Labelling these problems requires acknowledgement of absolute differences in social welfare, as well as the individual's perception.

ZC39 Health care system-related problem

Description

Problems related to the health care system.

Inclusion

person awaiting admission to elderly/nursing home ZC39.00
waiting period for investigation and treatment ZC39.01

Note

The diagnosis of problems with the health care system essentially requires the patient's expression of concern about them, with agreement about the existence of the problem and desire for help. Whatever the objective health care system, patients can consider this as a problem. Labelling these problems requires acknowledgement of absolute differences in the health care system as well as the individual's perception.

ZC90 Concern or fear of having a social problem

Description

Concern about or fear of having a social problem in a patient without a proven social problem.

Exclusion
If the patient has a social problem, code the social problem.

ZC99 Other specified social problems influencing health status

Inclusion
discrimination race/religion/gender ZC99.00
feeling lonely ZC99.01
problem illegal stay ZC99.02

Exclusion
air pollution AD45
all the other ZC classes

I INTERVENTIONS AND PROCESSES

Note
The principle in Interventions and processes is the core of health treatment closely related to quaternary prevention.

Quaternary prevention: action taken to protect individuals (persons or patients) from medical interventions that are likely to cause more harm than good.

-1 DIAGNOSTIC AND MONITORING INTERVENTIONS

Description
A clinical intervention intended to diagnose and monitor a patient's disease, condition or injury.

-101 Complete examination or health evaluation

Description
Complete examination of one body system or the whole body including mental and social problem-related examination, performed in own practice.

Exclusion
diagnostic questionnaires -111

-102 Partial examination or health evaluation

Description
An examination of a specific part of a body system or specific mental functions, or social problem related, performed in own practice.

Inclusion
auscultation
blood pressure measurement

body temperature measurement
dermatoscopy
diagnostic questionnaires
gynaecological internal examination
opthalmoscopy (fundoscopy)
oximetry
palpitation
pelvic examination
percussion
rectal examination
visual inspection

Exclusion
pregnancy care W309

-103 Sensitivity test

Description
Performing a test or requesting a test to detect/exclude allergy.

Inclusion
food sensitivity test
Mantoux test
methacholine challenge test
patch test
radioallergosorbent test (RAST) test
skin prick test

Exclusion
desensitisation -202

-104 Microbiological or immunological test

Description
Performing a test or requesting a test to detect/exclude microorganisms/immuno-logical mechanisms.

Inclusion
antibody test
CRP
cultures test
DNA/RNA test for the detection of the causative agent
HPV-DNA test
serological/immunological tests

-105 Blood test

Description
Performing a test or requesting a test for all determinations in blood.

Inclusion
blood group test
clinical chemistry tests in blood
coagulation tests
haematology tests
measurement of creatinine clearance

Exclusion
microbiological/serological and immunonological test in blood sample -104
sensitivity test in a blood sample (RAST/allergy) -103

-106 Urine test

Description
Performing a test or requesting a test for all determinations in urine.

Inclusion
albumin/creatinine ratio in urine

Exclusion
microbiological/serological and immunological test in urine sample -104
urine cytology -108

-107 Faeces test

Description
Performing a test or requesting a test for all determinations in faeces.

Inclusion
parasite faeces test

Exclusion
microbiological/serological and immunological test in faeces sample -104

-108 Histological and exfoliative cytology

Description
Performing a test or requesting a test to examine the structure of tissues and cells
under a (electronic)microscope.

Inclusion
anatomical pathology

biopsy of skin
histological or cytological examination of tissue or fluid retrieved by puncture or
    biopsy or excision or swabbing or collecting
urine cytology

Exclusion
semen analysis -109
sputum analysis -109
trichomonas vaginalis test -109

-109 Other specified laboratory test

Inclusion
CSF (cerebralspinal fluid) test
DNA/genetic/chromosome test
Helicobacter pylori breath test
pH fluorine test
semen analysis
sputum analysis without culture
sweat test
trichomonas vaginalis test

Exclusion
sputum culture -104

-110 Specific physical function test

Description
Measuring physical function of ear, eye, lungs, etc. using a specific device.

Inclusion
audiometry
spirometry
tonometry
tympanometry

Exclusion
electrical tracing tests as EKG/Holter -114
vision test; colour test; visual field test; calorimetric test; reflex test – all -102

-111 Standard mental, cognitive, physical functioning tests and questionnaires

Description
Performing a test or questionnaire or request for a test or questionnaire for assessment
of mental, cognitive or physical functioning.

Inclusion
anxiety test
dementia test
depression test
intelligence test

-112 Diagnostic endoscopy

Description
Performing a scopy inside the body by using an endoscope.

Inclusion
anoscopy
arthroscopy
bronchoscopy
colonoscopy
colposcopy
gastroscopy
hysteroscopy
laparoscopy
laryngoscopy
mediastinoscopy
pharyngoscopy
rectoscopy
rhinoscopy
sigmoidoscopy
tracheoscopy

Exclusion
dermatoscopy -102
fundoscopy -102
ophtalmoscopy -102

-113 Diagnostic imaging and radiology

Description
Diagnostic radiology refers to the field of medicine that uses non-invasive imaging scans for diagnosing a problem. The tests and equipment used sometimes involves low doses of radiation to create highly detailed images of an area. In some parts of the world the primary care physician has the possibility to do X-ray and ultrasound investigations in their own practice.

Inclusion
computerised tomography (CT) -113.00
magnetic resonance imaging (MRI) -113.01
ultrasound for foetal growth measurement
ultrasound of foetal structure

ultrasound imaging -113.02
X-ray -113.03

-114 Electrical tracing

Description
A test used to measure the electrical activity of an organ (e.g. heart, nerve, brain, muscle).

Inclusion
electrocardiogram
electroencephalogram (EEG)
electromyogram (EMG)
electronystagmography (ENG)
exercise electrocardiogram
Holter monitoring

-199 Other specified diagnostic interventions

Inclusion
diagnostic laparotomy
skin photo
tourniquet test

-2 THERAPEUTIC AND PREVENTIVE INTERVENTIONS

Description
The classes presented here are to be used for interventions performed by the provider him or herself.

-201 Pharmacotherapy and prescription

Inclusion
administration of medication
prescribing of injectable drug
prescribing of medication
renewal of medication

Exclusion
injection of medication with local effect -210
preventive immunisation/medication -202

-202 Preventive immunisation and medication

Inclusion
routine vaccination, children
vaccination

Coding hint

If contraceptive medication is prescribed for medical reasons, use the code for medication -201.

-203 Observation, health education, advice and diet

Description

Monitoring of health problems and advice on healthy behaviour.

Inclusion

advice on healthy behaviour
advice on prevention of health problems
advice on pregnancy and family planning
advice on prevention of injury
advice on prevention of violence
advice regarding the use of health services
advice regarding occupational health/social problems
monitoring of medication use
watchful waiting

Exclusion

therapeutic counselling/listening -212

-204 Incision, drainage, flushing, aspiration and removal body fluid

Inclusion

ascitic fluid puncture
incision of abscess
irrigation of ear/eye
paracentesis
puncture/aspiration of bursa
puncture/aspiration of cyst
puncture/aspiration of ganglion
puncture/aspiration of haematoma
puncture/aspiration of joint
puncture/aspiration of lungs
puncture/aspiration of urinary bladder

-205 Excision, removal of tissue, destruction, debridement and cauterisation

Inclusion

autolytic debridement
burning cauterisation
chemical cauterisation
chemical debridement
cold cauterisation

electric cauterisation
excision or removal of nail
excision or removal of tissue
extraction of tooth
laser cauterisation
mechanical debridement
removal of foreign body
surgical debridement

-206 Instrumentation, catheterisation, intubation and dilation

Inclusion
catheterisation
endotracheal intubation
enema
intravenous cannulation
lacrimal dilatation
tracheostomy
tympanostomy tube insertion

Exclusion
implantation of a hormone or long-acting drug -209
incision/drainage/flushing/aspiration/removal body fluid -204

-207 Repair-suture or cast

Description
Applying and removing cast, sutures, stitches, surgical glue and strip-plaster.

Inclusion
repair of perineum
repair of vulva
strip-plaster
surgical glue
suture/stitches

-208 Taping or strapping

Description
Application of adhesive bandages or tape (depending on the area), used to secure or
stabilise an injured or painful joint.

Inclusion
strapping for sprains
treatment of luxation or dislocation

-209 Application or removal of devices

Description
Any device intended to be used for medical purposes.

Inclusion
brace(s)
hernia support
insertion of an implant containing hormones or a long-acting drug
orthopaedic prosthetic(s)
orthose(s)
pacemaker
sling
vaginal pessary/IUD

-210 Local injection and infiltration

Description
Administering an injection for local effect.

Inclusion
bursa injection
intra-articular injection
sclerosing injection for varices
tendon sheath injection

-211 Dressing, pressure, compression and tamponade

Inclusion
application of eye pad
pressure bandage
tamponade (blockage to stop bleeding)
wound dressing

-212 Therapeutic counselling

Description
A process of consultation and discussion in which the provider (the counsellor) listens
and offers guidance or advice to the patient who is experiencing difficulties.

Inclusion
counselling for a specific disease
motivational interview
supportive psychotherapy

-215 Delivery-related Interventions

Inclusion

artificial rupture of the amniotic membranes
assisted vaginal delivery
delivery by caesarean section
episiotomy
external version of foetus
manual removal of retained placenta
medical induction of labour per orifice

Exclusion

pregnancy care W309
repair of perineum -207
repair of vulva -207

-299 Other specified treatment and therapeutic and preventive interventions

Inclusion

cardiopulmonary resuscitation
oxygen therapy
physical medicine/rehabilitation and acupuncture done in own practice, without a
    referral to another provider
uterine curettage

-3 PROGRAMMES RELATED TO REPORTED CONDITIONS

Description

These care programmes consist of a combination of various interventions such as asking questions during anamnesis, blood and urine tests, spirometry, advice and policy options, performed in primary care practice.

In general several health professionals are involved in a 'programme'. This implies that a care plan needs to reflect the integrated approach of all health professionals involved. This could also be referred to as the bio-psycho-social way of working and thinking.

Coding hint

In order to understand exactly what has been done in the context of the programme, the separate interventions in Component -2 should/must be coded.

Note

The programmes are directly connected to specific Chapters. The codes therefore contain the prefix of these chapters instead of a dash.

K301 Cardiovascular programme
K302 Heart failure programme
P303 Dementia (management) programme
P304 Depression (management) programme
P305 Other specified mental programme
R306 Asthma programme
R307 COPD programme
T308 Diabetes programme
W309 Pregnancy care

Inclusion
pregnancy check-up
pregnancy surveillance

A310 Polypharmacy care

Description
Personal health surveillance related to polypharmacy.

A350 Complex and integral care programme

Description
Integral care is an organising principle for care delivery with the aim of achieving improved patient care through better coordination of services provided. Integration is the combined set of methods, processes and models that seek to bring about this improved coordination of care. It is care that is planned with people who work together to understand the service user and their carer(s), puts them in control and coordinates and delivers services to achieve the best outcomes.

Exclusion
frailty elderly programme A351

A351 Frailty elderly programme

Description
Specific programme for frailty people. Frailty defines the group of older people who are at highest risk of adverse outcomes such as falls, disability, admission to hospital or the need for long-term care.

A352 Palliative care and end of life care

Description
Palliative care is an approach that improves the quality of life of patients and their families facing the problem associated with life-threatening illnesses, through the prevention and relief of suffering by means of early identification and impeccable assessment and treatment of pain and other problems, physical, psychosocial and spiritual.

X399 Other specified programmes related to reported conditions

-4 RESULTS

-401 Result of test or procedure requested by own provider

Description

Results from tests or procedures ordered/performed by the health care provider: blood, imaging, electrical tracing or other.

Note

This code can only be used to classify a reason for encounter.

-402 Result of an examination or test from another provider

Description

Results from tests or procedures ordered/performed by other health care providers.

Note

This code can only be used to classify a reason for encounter.

-5 CONSULTATION, REFERRAL AND OTHER REASONS FOR ENCOUNTER

-501 Encounter or problem initiated by provider

Description

The provider asks about a health problem that is not put forward by the patient.

Inclusion

problem managed by the provider, that was not on the patient's agenda

Exclusion

encounter/problem initiated by other than patient/provider -502

Note

This code can only be used to classify a reason for encounter.

-502 Encounter or problem initiated by other than patient or provider

Description

Encounter requested by a third party.

Exclusion

encounter/problem initiated by provider -501

Note

This code can only be used to classify a reason for encounter. If the patient is unable to state the reason for encounter, use the reason stated by the accompanying person.

-503 Consultation with primary care provider

Inclusion
telemedicine consultation with primary care provider

Exclusion
consultation with specialist -504
referral to another provider -505
referral to physician/specialist/clinic/hospital -506
other referral -599

Note
Treatment responsibility remains with the original primary care provider.

-504 Consultation with specialist

Inclusion
telemedicine consultation with specialist

Exclusion
consultation with primary care provider -503
other referral -599
referral to another provider -505
referral to physician/specialist/clinic/hospital -506

Note
Treatment responsibility remains with the original primary care provider.

-505 Referral to other primary care provider

Inclusion
referral to chiropodist
referral to chiropractor
referral to dentist
referral to dietician
referral to home health worker
referral to midwife
referral to occupational therapist
referral to orthodontist
referral to optician
referral to other GP or FP
referral to psychologist
referral to physiotherapist
referral to nurse
referral to social worker

Exclusion
referral to specialist -506
referral to institution for rehabilitation -599

-506 Referral to specialist, clinic or hospital

Inclusion
referral to specialist
referral to disease-specific out-/inpatient clinics

Exclusion
referral to institution for rehabilitation -599

-599 Other specified consultations, referrals and reasons for encounter

Inclusion
advice to contact a service outside the regular health service (e.g. patient associations, unemployment services)
referral to a nursing home or hospice
referral to a service for rehabilitation

Exclusion
referral to other provider, nurse, therapist, social worker -505
referral to specialist, clinic or hospital -506

Note
This code can only be used to classify a reason for encounter. If the patient is unable to state the reason for encounter, use the reason stated by the accompanying person.

-6 ADMINISTRATIVE
-601 Administrative procedure

Description
This code is designed to classify those instances where provision of a written document or form by the provider for the patient or agency is warranted by existing regulations, laws or customs.

Inclusion
billing issues
certificates (e.g. sick leave/driver's licence/death)
filling in documents or forms
health record issues
request for information

Exclusion

medical examination/health evaluation complete -101
medical examination/health evaluation partial -102
standard mental/cognitive/physical functioning tests and questionnaires -111

-602 Formulation of plan for care, management, treatment or intervention

Exclusion

execution of programmes related to reported conditions -3

II FUNCTIONING AND FUNCTIONING RELATED

Description

This chapter allows for the description of Functioning and Functioning related aspects of all persons (first and follow-up) contacts with the health care system in primary and community care settings. The Functioning and Functioning Related items are a selected subset of items from the WHO International Classification of Functioning, Disability and Health (ICF), which provides an overview of a person in a person-in-context approach, at a certain moment in time.

Where indicated in the references of the classes, a specific set of items is available in the form of self-administered tools for the assessment of functioning (and disability). These sets can be regarded as implementations of ICF within a specific use case.

In the first instance there is the World Health Organization Disability Assessment Schedule 2.0 (WHODAS 2.0) from WHO, which is available at www.psychiatry.org/dsm5.

- The WHODAS 2.0 is a general tool for the assessment of difficulties due to health/mental health conditions. This assessment tool is advised to be used for the collection of disability data for adults aged 18 years and older.
- For specific use in primary health care settings the Primary Care Functioning Scale (PCFS) has been developed with an intended population age group 50+ with multimorbidity. The PCFS needs further testing.
- In addition, the 'Arrèts de Travail en médecine générale à partir de la Classification Internationale de Fonctionncment' (ATCIF) has been developed for sick-leave prescription. In many countries sick-leave prescriptions are frequently used in primary health care/general practices. Using the ICF for sick-leave prescription, instead of the traditional medical approach, supports and changes the way health professionals and patients communicate in the work-related context.

The questions from these questionnaires have been itemised in Chapter II, and their use is encouraged whenever relevant, as separate items or scored with the WHODAS 2.0, the PCFS or the ATCIF.

If greater detail on Functioning and Functioning related aspects is required than that available within the presented selection of items, the WHO ICF should be consulted.

Access to ICF classification: http://apps.who.int/classifications/icfbrowser/

2F FUNCTIONING

Description

Functioning of a person can be defined by the complexity of components such as the physiological functions of body systems and psychological functions, anatomical features of parts of the body such as organs, limbs and their components and the execution of tasks or actions by an individual as such or the involvement of a person in a life situation.

Physiological functions of body systems and psychological functions are referred to as body functions (body and body system level).

Anatomical features of parts of the body such as organs, limbs and their components are referred to as body structures (body level). Not as such classified in the ICPC-3. In the ICPC-3 anatomical terms are harmonised with the Foundational Model of Anatomy, like the ICD-11.

Execution of tasks or actions by an individual are referred to as Activities (person level).

The involvement of a person in a life situation is referred to as Participation (person in social context level).

From the primary health care point of view, activities and participation are the core part for shaping a person-centred approach. This means that in the ICPC-3 the Activities and participation chapter comes first, followed by the Functions chapter.

2F0 Activities and participation

Description

Execution of tasks or actions by an individual are referred to as Activities (person level).

The involvement of a person in a life situation is referred to as Participation (person in social context level).

2F01 Watching

Description

Using the sense of seeing intentionally to experience visual stimuli.

Inclusion

visually tracking an object
watching a sporting event
watching people
watching children playing

2F02 Listening

Description
Using the sense of hearing intentionally to experience auditory stimuli.

Inclusion
listening to a radio
listening to the human voice
listening to music
listening to a lecture
listening to a story told

2F03 Basic learning

Description
Basic learning is a broad concept for developing competencies.

Inclusion
actions with objects
acquiring concepts and information
acquiring language
acquiring skills
imitating or mimicking others
learning to read
learning to write
learning to calculate
rehearsing

2F04 Focusing attention

Description
Intentionally focusing on specific stimuli.

Inclusion
filtering out distracting noises

2F05 Thinking

Description
Formulating and manipulating ideas, concepts and images, whether goal-oriented or not, either alone or with others.

Inclusion
creating fiction
proving a theorem
playing with ideas
brainstorming

meditating
pondering
speculating
reflecting

2F06 Reading

Description
Performing activities involved in the comprehension and interpretation of written language (e.g. books, instructions or newspapers in text or Braille), for the purpose of obtaining general knowledge or specific information.

2F07 Calculating

Description
Performing computations by applying mathematical principles to solve problems that are described in words and producing or displaying the results.

Inclusion
computing the sum of three numbers
finding the result of dividing one number by another

2F08 Solving problems

Description
Finding solutions to questions or situations by identifying and analysing issues, developing options and solutions, evaluating potential effects of solutions and executing a chosen solution.

Inclusion
resolving a dispute between two people

2F09 Making decisions

Description
Making a choice among options, implementing the choice and evaluating the effects of the choices that need to be done.

Inclusion
deciding to undertake a task
selecting and purchasing a specific item
undertaking one task from among several tasks

2F10 Undertaking a single task

Description
Carrying out simple or complex and coordinated actions related to the mental and physical components of a single task.

Inclusion

carrying out, completing and sustaining a task
initiating a task
organising time, space and materials for a task
pacing task performance

2F11 Undertaking multiple tasks

Description

Carrying out simple or complex and coordinated actions as components of multiple, integrated and complex tasks in sequence or simultaneously (ICF).

2F12 Carrying out daily routine

Description

Carrying out simple or complex and coordinated actions in order to plan, manage and complete the requirements of day-to-day procedures or duties.

Inclusion

budgeting time
making plans for separate activities throughout the day

2F13 Handling stress

Description

Carrying out simple or complex and coordinated actions to cope with pressure, emergencies or stress associated with task performance (ICF).

Inclusion

coping with emergencies
coping with pressure
coping with stress

2F14 Communicating with – receiving – spoken messages

Description

Comprehending literal and implied meanings of messages in spoken language.

2F15 Speaking

Description

Producing words, phrases and longer passages in spoken messages with literal and implied meaning.

Inclusion

expressing a fact
telling a story in oral language

2F16 Conversing

Description
Starting, sustaining and ending an interchange of thoughts and ideas, carried out by means of spoken, written, signed or other forms of language, with one or more people one knows or who are strangers, in formal or casual settings (ICF).

2F17 Discussing

Description
Starting, sustaining and ending an examination of a matter, with arguments for or against, or debate carried out by means of spoken, written, sign or other forms of language, with one or more people one knows or who are strangers, in formal or casual settings (ICF).

2F18 Using communication devices and techniques

Description
Using devices, techniques and other means for the purposes of communicating.

Inclusion
calling a friend on the telephone

2F20 Changing basic body position

Description
Getting into and out of a body position and moving from one location to another.

Inclusion
getting into and out of position of sitting
getting into and out of position of standing
getting into and out of position of kneeling
getting into and out of position of squatting
getting up out of a chair to lie down on a bed

2F21 Maintaining a body position

Description
Staying in the same body position as required, such as remaining seated or remaining standing for carrying out a task, in play, work or school (ICF).

2F22 Transferring oneself

Description
Moving from one surface to another without changing body position.

Inclusion
moving from a bed to a chair
sliding along a bench

2F23 Lifting and carrying object

Description
Raising up an object or taking something from one place to another.

Inclusion
carrying a box
carrying a child from one room to another
lifting a cup
lifting a toy

2F25 Fine hand use

Description
Performing the coordinated actions of handling objects such as required to lift coins off a table or turn a dial or knob.

Inclusion
handling objects
picking up objects using one's hand, fingers and thumb
manipulating objects using one's hand, fingers and thumb
releasing objects using one's hand, fingers and thumb

2F26 Hand and arm use

Description
Performing the coordinated actions required to move objects with hands and arms.

Inclusion
manipulating objects by using hands and arms
moving objects by using hands and arms
throwing or catching an object
turning door handles

2F27 Walking long distances and short distances

Description
Walking for more or less than a kilometre, such as walking around rooms or hallways, within a building or for short distances outside, or walking for more than a kilometre, such as across a village or town, between villages or across open areas (ICF).

2F28 Climbing (steps)

Description
Moving the whole body upwards or downwards, over surfaces or objects.

Inclusion
climbing curbs
climbing ladders
climbing rocks
climbing stairs
climbing steps

2F29 Moving around within the home

Description
Walking and moving around in one's home, within a room, between rooms and around the whole residence or living area (ICF).

2F30 Moving around outside the home and other buildings

Description
Walking and moving around close to or far from one's home and other buildings, without the use of transportation, public or private, such as walking for short or long distances around a town or village (ICF).

2F31 Moving around using equipment

Description
Moving the whole body from place to place, on any surface or space, by using specific devices designed to facilitate moving or create other ways of moving around.

Inclusion
using a walker
using scuba equipment
using skates
using skis
using a walking stick
using a wheelchair

2F32 Using transportation

Description
Using transportation to move around as a passenger.

Inclusion
being driven in a boat
being driven in a bus

being driven in a car
being driven in a jitney
being driven in a pram
being driven in a private or public taxi
being driven in a stroller
being driven in a rickshaw
being driven in a train
being driven in a tram
being driven in a wheelchair
being driven in an aircraft
being driven in an animal-powered vehicle
being driven by subway
using humans for transportation

2F33 Driving

Description
Being in control of and moving a vehicle or the animal that draws it, travelling under one's own direction or having at one's disposal any form of transportation appropriate for age.

Inclusion
driving a bicycle
driving a boat
driving a car
driving a motorcycle
driving an animal-powered vehicle

2F34 Washing oneself

Description
Washing and drying one's whole body, or body parts, using water and appropriate cleaning and drying materials or methods.

Inclusion
bathing
showering
washing hands and feet
washing face and hair
drying with a towel

2F35 Caring for body parts

Description
Looking after those parts of the body that require more than washing and drying.

Inclusion
looking after genitals
looking after face
looking after nails
looking after scalp
looking after skin
looking after teeth

2F36 Toileting

Description
Planning and carrying out the elimination of human waste and cleaning oneself afterwards.

Inclusion
carrying out the elimination of human waste of defaecation
carrying out the elimination of human waste of menstruation
carrying out the elimination of human waste of urination
cleaning oneself after defaecation
cleaning oneself after menstruation
cleaning oneself after urination

2F37 Dressing

Description
Carrying out the coordinated actions and tasks of putting on and taking off clothes and footwear in sequence and in keeping with climatic and social conditions.

Inclusion
putting on and taking off clothes and footwear in correct sequence
putting on, adjusting and removing a shirt
putting on, adjusting and removing a skirt
putting on, adjusting and removing a blouse
putting on, adjusting and removing pants
putting on, adjusting and removing undergarments
putting on, adjusting and removing a sari
putting on, adjusting and removing a kimono
putting on, adjusting and removing tights
putting on, adjusting and removing a hat
putting on, adjusting and removing gloves
putting on, adjusting and removing a coat
putting on, adjusting and removing shoes
putting on, adjusting and removing boots
putting on, adjusting and removing sandals
putting on, adjusting and removing slippers

2F38 Eating

Description
Carrying out the coordinated tasks and actions of eating food that has been served, bringing it to the mouth and consuming it in culturally acceptable ways, cutting or breaking food into pieces, opening containers and packets, using eating implements, having meals, feasting or dining (ICF).

2F39 Drinking

Description
Taking hold of a drink, bringing it to the mouth and consuming the drink in culturally acceptable ways, mixing, stirring and pouring liquids for drinking, opening bottles and cans.

Inclusion
drinking from a breast
drinking running water from a tap
drinking running water from a spring
drinking through a straw

2F40 Looking after one's health

Description
Ensuring physical comfort, health and physical and mental well-being.

Inclusion
avoiding harms to health
following safe sex practices
getting immunisations
getting regular physical examinations
keeping warm or cool
maintaining a balanced diet
maintaining an appropriate level of physical activity
using condoms

Coding hint
In case a patient indicates experiencing a problem in managing one's lifestyle related to specified habits, code to AP40 Problems related to lifestyle, or PS13, PS14, PS15, PS16, TD66, TS51.

Note
The inclusions in this class are intended for general registration purposes to be informed about the person's health-related habits.

2F45 Doing housework

Description

Managing a household by cleaning the house, washing clothes, using household appliances, storing food and disposing of garbage, such as by sweeping, mopping, washing counters, walls and other surfaces; collecting and disposing of household garbage; tidying rooms, closets and drawers; collecting, washing, drying, folding and ironing clothes; cleaning footwear; using brooms, brushes and vacuum cleaners; using washing machines, dryers and irons (ICF).

2F46 Assisting others

Description

Assisting household members and others with their learning, communicating, self-care, movement, within the house or outside; being concerned about the well-being of household members and others (ICF).

Inclusion

assisting others with self-care
assisting others in movement
assisting others in communication
assisting others in interpersonal relations
assisting others in nutrition
assisting others in health maintenance

2F49 Basic interpersonal interactions

Description

Interacting with people in a contextually and socially appropriate manner.

Inclusion

responding to the feelings of others
showing consideration and esteem when appropriate

2F50 Complex interpersonal interactions

Description

Maintaining and managing interactions with other people, in a contextually and socially appropriate manner, when, for example, playing, studying or working with others.

Inclusion

acting in accordance with social rules and conventions
acting independently in social interactions
controlling verbal and physical aggression
regulating emotions and impulses

2F51 Relating with strangers

Description
Engaging in temporary contacts and links with strangers for specific purposes.

Inclusion
asking for directions
asking for information
making a purchase

2F52 Formal relationships

Description
Creating and maintaining specific relationships in formal settings.

2F53 Informal social relationships

Description
Entering into relationships with others, such as casual relationships with people living in the same community or residence, or with co-workers, students, playmates, people with similar backgrounds or professions (ICF).

2F54 Family relationships

Description
Creating and maintaining kinship relationships, such as those with members of the nuclear family, extended family, foster and adopted family and step-relationships, more distant relationships such as second cousins or legal guardians (ICF).

2F55 Intimate relationships

Description
Creating and maintaining close or romantic relationships between individuals.

Inclusion
maintaining a close relationship between husband and wife
maintaining a close relationship between lovers
maintaining a close relationship between sexual partners

2F56 Education and school

Description
Gaining admission to school, higher education and vocational training, engaging in all school-related responsibilities and privileges, and learning the course material, subjects and other curriculum requirements in all education programmes.

Inclusion
attending school regularly
working cooperatively with other students
taking directions from teachers
organising, studying and completing assigned tasks and projects
advancing to other stages of education
higher education
school education
vocational training

2F57 Acquiring, keeping and terminating a job

Description
Seeking, finding and choosing employment, being hired and accepting employment, maintaining and advancing through a job, trade, occupation or profession, and leaving a job in an appropriate manner (ICF).

2F58 Remunerative employment

Description
Engaging in all aspects of work, as an occupation, trade, profession or other form of employment, for payment, as an employee, full or part time, or self-employed, such as seeking employment and getting a job, doing the required tasks of the job, attending work on time as required, supervising other workers or being supervised, and performing required tasks alone or in groups (ICF).

Inclusion
working full time
working part time

2F59 Non-remunerative employment

Description
Engaging in all aspects of work in which pay is not provided, full time or part time, including organised work activities, doing the required tasks of the job, attending work on time as required, supervising other workers or being supervised, and performing required tasks alone or in groups.

Inclusion
doing charity work
doing volunteer work
working for a community or religious group without remuneration
working around the home without remuneration

2F60 Community life

Description
Engaging in aspects of community social life, such as engaging in charitable organisations, services clubs or professional social organisations (ICF).

2F61 Recreation and leisure

Description
Engaging in any form of play, recreational or leisure activity.

Inclusion
engaging in crafts or hobbies
engaging in informal or organised play and sports
engaging in programmes of physical fitness
engaging in relaxation, amusement or diversion
going to art galleries, museums, cinemas or theatres
playing musical instruments
reading for enjoyment
sightseeing, tourism and travelling for pleasure

2F69 Other specified activities and participation

Description
For other specified activities and participation, not presented in this section, please consult the ICF for more detail.

2F7 Functions

Description
Physiological functions of body systems and psychological functions are referred to as body functions (body and body system levels).

Note
In the subcomponent Functions, the classes and codes can be used to assess the 'problem level', i.e. the level of impairment of the specified function, not to address a RFE or an episode of care. Describing the nature and assessing the severity of the problem offers the possibility of the follow-up of care and addressing changes over time, such as a decrease or an increase of the impairment/problem.

Some of the class names in Functions overlap with class names in the Symptoms, complaints and abnormal findings component. These class names, such as 'dizziness', refer to the same phenomenon, but serve a different purpose or role.

E.g. 'Dizziness' as an impairment (problem in a function – 2F83) can be used in a descriptive way in order to understand to what extent a person experiences dizziness as a problem. Without coding the (level of) impairment, the dizziness is just a textual element, that is difficult to trace. Coding as a Function makes the dizziness, and the changes in it, traceable, available for discussion and countable.

Symptoms, complaints and abnormal findings are to be classified and coded at the level of the relevant body system chapters. In a situation where a person expresses that they experience 'dizziness' as the RFE, the class/code to be used for dizziness is in Neurological system – NS09 Vertigo or dizziness, with no further possibility for expression of detail. There it is meant to classify the symptom, complaint or abnormal finding.

2F71 Energy level

Description
Mental and physical functions that produce vigour and stamina (ICF).

2F72 Sleep functions

Description
General mental functions of periodic, reversible and selective physical and mental disengagement from one's immediate environment accompanied by characteristic physiological changes (ICF).

2F73 Attention functions

Description
Specific mental functions of focusing on an external stimulus or internal experience for the required period of time (ICF).

2F74 Memory functions

Description
Specific mental functions of registering and storing information and retrieving it as needed (ICF).

2F75 Emotional functions

Description
Specific mental functions related to the feeling and affective components of the processes of the mind (ICF).

2F80 Seeing functions

Description
Sensory functions relating to sensing the presence of light and sensing the form, size, shape and colour of the visual stimuli (ICF).

2F81 Hearing functions

Description
Sensory functions relating to sensing the presence of sounds and discriminating the location, pitch, loudness and quality of sounds (ICF).

2F82 Balance

Description
Sensory functions of the inner ear related to determining the balance of the body (ICF).

2F83 Dizziness

Description
Sensation of motion involving either oneself or one's environment.

Inclusion
sensation of rotating
sensation of swaying
sensation of tilting

2F84 Pain functions

Description
Sensation of unpleasant feeling indicating potential or actual damage to some body structure (ICF).

2F85 Exercise tolerance functions

Description
Functions related to respiratory and cardiovascular capacity as required for enduring physical exertion (ICF).

2F86 Sexual functions

Description
Mental and physical functions related to the sexual act, including the arousal, preparatory, orgasmic and resolution stages (ICF).

Inclusion
functions of sexual arousal
preparatory, orgasmic and resolution phase
functions related to sexual interest
sexual performance
penile erection
clitoral erection
vaginal lubrication

ejaculation
orgasm

2F90 Mobility of joint functions

Description
Functions related to the range and ease of movement of a joint (ICF).

2F91 Muscle power functions

Description
Functions related to the force generated by the contraction of a muscle or muscle groups (ICF).

2F99 Other specified functions

Description
For other specified functions, not presented in this section, please consult the ICF for more detail.

2R FUNCTIONING RELATED

Description
Functioning related factors describe the context in which functioning takes place and how functioning is executed. They are made up by the environmental factors the person lives in (the things outside the person) and the personal characteristics in which one person differs from another person.

2R0 Environmental factors

Description
Environmental factors are made up of the environment the person lives in (the things outside the person).

2R01 Food

Description
Any natural or human-made object or substance gathered, processed or manufactured to be consumed.

Inclusion
breast milk
herbs
liquids of different consistencies
minerals (vitamin and other supplements)
prepared food

processed food
raw food

2R02 Drinking water

Description
Water suitable and safe for personal consumption.

2R03 Drugs (medication)

Description
Any natural or human-made object or substance gathered, processed or manufactured for medicinal purposes.

Inclusion
allopathic medication
naturopathic medication

2R04 Housing

Description
The availability of a house or shelter for persons to live in.

Inclusion
shelter

2R05 Sanitation

Description
The availability of, or access to, means for safe water for drinking and washing, and adequate treatment and disposal of human excreta and sewerage.

2R06 Assistive products and technology for personal indoor and outdoor mobility and transportation

Description
Adapted or specially designed equipment, products and technologies that assist people to move inside and outside buildings.

Inclusion
adaptations to vehicles
scooters
special cars and vans
transfer devices
walking devices (such as canes or crutches)
wheelchairs

2R07 Natural environment and human-made changes to environment

Description
This class is about animate and inanimate elements of the natural or physical environment, and components of that environment that have been modified by people, as well as characteristics of human populations within that environment (ICF).

2R08 Immediate family

Description
Individuals related by birth, marriage or other relationship recognised by the culture as immediate family.

Inclusion
adoptive parents
children
foster parents
grandparents
parents
partners
siblings
spouses
support by immediate family

2R09 Friends

Description
Individuals who are close and ongoing participants in relationships characterised by trust and mutual support (ICF).

Inclusion
support by friends

2R10 Acquaintances, peers, colleagues, neighbours and community members

Description
Individuals who are familiar to each other as acquaintances, peers, colleagues, neighbours and community members, in situations of work, school, recreation or other aspects of life and who share demographic features such as age, gender, religious creed or ethnicity or pursue common interests (ICF).

Inclusion
support by acquaintances
support by peers
support by colleagues
support by neighbours
support by community members

2R16 Health professionals

Description
All service providers working within the context of the health system.

Inclusion
audiologists
doctors
medical social workers
medical specialists
nurses
occupational therapists
orthotist-prosthetists
physiotherapists
speech therapists

2R17 Individual attitudes of immediate family members

Description
General or specific opinions and beliefs of immediate family members about the person or about other matters (e.g. social, political and economic issues) that influence individual behaviour and actions (ICF).

2R18 Individual attitudes of health professionals

Description
General or specific opinions and beliefs of health professionals about the person or about other matters (e.g. social, political and economic issues) that influence individual behaviour and actions (ICF).

2R19 Social security

Description
Services, systems and policies aimed at providing income support to people who, because of age, poverty, unemployment, health condition or disability, require public assistance that is funded either by general tax revenues or contributory schemes (ICF).

2R20 Home health services

Description
Individuals who provide services to support individuals in their daily activities and maintenance of performance at work, education or other life situation, provided either through public or private funds, or else on a voluntary basis.

Inclusion
Nannies
paid help
personal assistants
primary caregivers
providers of support for home-making and maintenance
transport assistants

2R29 Other specified external factors

Description
For other specified external factors, not presented in this section, it is advised to consult the ICF for more detail.

2R3 PERSONALITY FUNCTIONS

Description
Personality functions are personal characteristics in which one person differs from another person.

Personality functions require the persons own perception and expression of, and to what extent a personal characteristic plays a role in, the context of the person's health.

Personality functions should only be used if provided by the person her- or himself and with consent for use or re-use. It is not to express the health provider's opinion about the person.

2R30 Extraversion

Description
Mental functions that produce a personal disposition that is outgoing, sociable and demonstrative.

Inclusion
being demonstrative
being outgoing
being sociable

2R31 Agreeableness

Description
Mental functions that produce a personal disposition that is cooperative, amicable and accommodating.

Inclusion
being accommodating
being amicable
being cooperative

2R32 Conscientiousness

Description
Mental functions that produce a personal disposition such as in being hard-working, methodical and scrupulous.

Inclusion
being hard-working
being methodical
being scrupulous

2R33 Psychic stability

Description
Mental functions that produce a personal disposition that is even-tempered, calm and composed.

Inclusion
being calm
being composed
being even-tempered

2R34 Openness to experience

Description
Mental functions that produce a personal disposition that is curious, imaginative, inquisitive and experience-seeking.

Inclusion
being curious
being experience-seeking
being imaginative
being inquisitive

2R35 Optimism

Description
Mental functions that produce a personal disposition that is cheerful, buoyant and hopeful.

Inclusion
being buoyant
being cheerful
being hopeful

2R36 Confidence

Description
Mental functions that produce a personal disposition that is self-assured, bold and assertive.

Inclusion
being assertive
being bold
being self-assured

2R37 Trustworthiness

Description
Mental functions that produce a personal disposition that is dependable and principled.

Inclusion
being dependable
being principled

2R39 Other specified Personality functions

Description
For other specified personality functions, not presented in this section, please consult the ICF for more detail.

IV EMERGENCY CODES

Description
The EM-codes are for emergency use with epidemiological importance for risk of (national or international) spreading of infections.

EM01 Code for emergency use
EM02 Code for emergency use
EM03 Code for emergency use
EM04 Code for emergency use
EM05 Code for emergency use
EM06 Code for emergency use
EM07 Code for emergency use
EM08 Code for emergency use
EM09 Code for emergency use

V EXTENSION CODES

Description
Extension codes are provided as supplementary codes or additional positions to give more detail or meaning to the initial code, if so desired. The Extension codes are not to be used without an initial code.

SV SCALE VALUE
PSV Problem Scale Value

Description

In the ICPC-3 no distinction is made between having a problem with a function or an activity or participation. For the Functioning components, the scale values are expressed in terms of the value level of the problem. Using these values at a certain point in time or over a period of time informs about actual Functioning situation or gives a 'snapshot' of the person. The values can also be used for goal setting.

The correspondence between the ICPC-3 and the severity scales (qualifiers) from the ICF is as follows:

PSV.0 NO problem	xxx.0 NO impairment/difficulty
PSV.1 MILD/MODERATE problem	xxx.1 MILD impairment/difficulty and xxx.2 MODERATE impairment/difficulty
PSV.2 SEVERE problem	xxx.3 SEVERE impairment/difficulty
PSV.3 COMPLETE problem	xxx.4 COMPLETE impairment/difficulty
PSV.9 NOT applicable	xxx.9 not applicable

In daily practice, health professionals and patients or clients find it difficult to differentiate between MILD or MODERATE. For this reason, MILD and MODERATE are merged into one value.

PSV.0 NO problem

There is no problem. The problem is absent or experienced as negligible.

PSV.1 MILD/MODERATE problem

The problem is experienced as slight, low, medium or fair.

PSV.2 SEVERE problem

The problem is experienced as high or extreme.

PSV.3 COMPLETE problem

The problem is experienced as total or complete.

PSV.9 NOT applicable

FBV Facilitator or Barrier value

Description

The correspondence between the Facilitator or Barrier values from the ICPC-3 and the barrier or facilitator from the ICF is as follows:

FBV.0 NO facilitator/NO barrier	xxx+0 NO facilitator and xxx.0 NO barrier
FBV.1 FULL facilitator	xxx+4 COMPLETE facilitator
FBV.2 STRONG facilitator	xxx+3 SUBSTANTIAL facilitator
FBV.3 MODERATE/MILD facilitator	xxx+2 MODERATE facilitator and xxx+1 MILD facilitator
FBV.4 MILD/MODERATE barrier	xxx.1 MILD barrier and xxx.2 MODERATE barrier
FBV.5 STRONG barrier	xxx.3 SEVERE barrier
FBV.6 FULL barrier	xxx.4 COMPLETE barrier
FBV.9 NOT applicable	xxx.9 not applicable

In daily practice, health professionals and patients or clients find it difficult to differentiate between MILD or MODERATE. For this purpose MILD and MODERATE are merged into one value.

FBV.0 NO facilitator/NO barrier
FBV.1 FULL facilitator
FBV.2 STRONG facilitator
FBV.3 MODERATE/MILD facilitator
FBV.4 MILD/MODERATE barrier
FBV.5 STRONG barrier
FBV.6 FULL barrier
FBV.9 NOT applicable

CSV Consent Scale Value

Description
The Consent Scale Value (CSV) is used by a patient or client to express the level of agreement concerning Personality functions (2R3). Without these values, the Personality functions have no specific meaning.

CSV.2+ COMPLETELY agree
CSV.1+ MODERATELY agree
CSV.0 NEUTRAL
CSV.1 MODERATELY disagree
CSV.2 COMPLETELY disagree

FEV Forced Expiratory Volume

Description
Forced Expiratory Volume (FEV) is a calculated ratio for the indication of the volume of air exhaled under forced conditions in the first second of expiration (FEV1). It is also called the person's vital capacity in persons with Chronic Obstructive Lung Disease.

The GOLD criteria or severity scale was developed by the Global Initiative for Chronic Obstructive Lung Disease.

GOL.1 GOLD 1 = mild: FEV1 is more than or equal to 80% predicted
GOL.2 GOLD 2 = moderate: between 50% to 80% FEV1 predicted
GOL.3 GOLD 3 = severe: between 30% to 50% FEV1 predicted
GOL.4GOLD 4 = very severe: less than 30% FEV1 predicted
GOL.5 GOLD not specified

NYHA New York Heart Association Functional Classification

NYH.1 NYHA Class I
No symptoms and no limitations in ordinary physical activity; e.g. shortness of breath when walking, climbing stairs, etc.

NYH.2 NYHA Class II
Mild symptoms (mild shortness of breath and/or angina) and slight limitations during ordinary activity.

NYH.3 NYHA Class III
Marked limitation in activity due to symptoms, even during 'less-than-ordinary activity' e.g. walking short distances (20–100 metres). Comfortable only at rest.

NYH.4 NYHA Class IV
Severe limitations. Experiences symptoms even while at rest. Mostly bedbound patients.

NYH.9 NYHA Class IX
No NYHA class listed or unable to determine.

TEM Temporality

COU.0 Subacute
COU.1 Acute
COU.2 Chronic

CAU Causality

These Class attributes are provided here for informative purposes only to address the causality of classes within a component. A number of these class attributes have been assigned with a specific colour, which is shown in the classification browser. The colouring is also used for the desk version to increase the informative value of the sheet.

CAU.0 Congenital
CAU.1 Hereditary
CAU.2 Infectious
CAU.3 Neoplasm
CAU.4 Injury
CAU.5 Lifestyle
CAU.6 Immunology
CAU.8 Other
CAU.9 Unknown

CHAPTER 13

Conversion from ICPC-3 to ICPC-2 and ICPC-1

ICPC-3	ICPC-2	ICPC-1
AF01	A98	A97
AF02	W11	W11
AF03	W12	W12
AF04	W10	W10
AF05	W14, Y14	W14, Y14
AF06	W13, Y13	W13, Y13
AG01	A97	A97
AG02	A97	A97
AG03	A97	A97
AG04	A97	A97
AG99	--	--
AI01	--	--
AI02	--	--
AI03	A20	A20
AI99	--	--
AP01	A98	A97
AP10	A98	A97
AP20	A98	A97
AP21	A98	A97
AP22	--	--
AP40	A99	A97
AP45	A98, P09	A97, P09
AP50	A23	--
AP60	A21, A23, A99, K22	A99
AP65	A21, A23, K22	--
AP70	--	--

DOI:10.1201/9781003197157-13

ICPC-3	ICPC-2	ICPC-1
AP80	A99	A99
AP99	A98	A97
AQ01	--	--
AQ02	--	--
AQ03	--	--
AQ04	--	--
AQ99	--	--
AR01	--	--
AR02	--	--
AR03	--	--
AR99	A99	A99
AS01	A01	A01
AS02	A02	A02
AS03	A03	A03
AS04	A04	A04
AS05	A04	A04
AS06	A05	A05
AS07	A06	A06
AS09	A08	A08
AS10	A09	A09
AS11	A10	A10
AS12	A11	L04
AS13	A16	A15, A16, A17
AS14	A29	A29
AS50	A91	A91, B85
AS52	A99	A99
AS53	A07	A07
AS90	A25, A26, A27	A25, A26, A27
AS91	A13	A13
AS92	A18, W21	--
AS99	A29	A29
AD01	A71	A71
AD02	A72	A72
AD03	A74	A74
AD04	A75	A75

ICPC-3	ICPC-2	ICPC-1
AD13	A76	A76
AD14	A77	A77
AD15	A70	A70, R70
AD16	A73	A73
AD17	A78	A78
AD23	A78, U71	A78, U71
AD24	A78	A78, A92
AD25	A79	A79
AD26	A99	A99
AD35	A81	A81
AD36	A80	A80
AD37	A82	A82
AD40	A84	A84
AD41	A85	A85
AD42	A87	A87
AD43	A89	A89
AD44	A86	A86
AD45	A88	A88
AD46	A92	A12
AD55	A90	A90
AD65	A93	A93
AD66	A94	A94
AD95	A95	A95
AD96	A96	A96
AD99	A99	A99
BS01	B02	B02, B03
BS50	B87	B87
BS51	B84, B99	B84, B86
BS52	B99	B99
BS90	B25, B26, B27	B25, B26, B27
BS99	B04, B29	B29, B04
BD01	B70	B70
BD02	B71	B71
BD03	B90	B90
BD04	B90	B90

ICPC-3	ICPC-2	ICPC-1
BD25	B72, B73, B74	B72, B73, B74
BD26	B75	B75
BD35	B77, B76	B77, B76
BD55	B79	B79
BD65	B78	B78
BD66	B80	B80
BD67	B81	B81
BD77	B82	B82
BD78	B83	B83
BD99	B99	B99
DS01	D01	A14, D01
DS02	D02	D02
DS03	D03	D03
DS04	D04	D04
DS05	D05	D05
DS06	D06	D06
DS07	D07	D02
DS08	D08	D08
DS09	D09	D09
DS10	D10	D10
DS11	D11	D11
DS12	D12	D12
DS13	D13	D13
DS14	D14	D14
DS15	D15	D15
DS16	D16	D16
DS17	D17	D17
DS18	D18	D18
DS19	D19, D29	D19, D29
DS20	D20	D20
DS21	D21	D21
DS50	D23	D96
DS51	D24, D25, D29	D24, D25, D29
DS90	D26, D27	D26, D27
DS99	D29	D29

ICPC-3	ICPC-2	ICPC-1
DD01	D70	D70
DD02	D71	D71
DD03	D72	D72
DD05	D73	D73
DD06	D95	D95
DD07	D96	D22
DD25	D74	D74
DD26	D75	D75
DD27	D76	D76
DD28	D77	D77
DD29	D78	D78
DD35	D80	D80
DD36	D79	D79
DD55	D81	D81
DD65	D82	D82
DD66	D83	D83
DD67	D84	D84
DD68	D84	D84
DD69	D85	D85
DD70	D86	D86
DD71	D87	D87
DD72	D88	D88
DD73	D89	D89
DD74	D90	D90
DD75	D91	D91
DD76	D91	D91
DD77	D92	D92
DD78	D93	D93
DD79	D94	D94
DD81	D97	D97
DD82	D98	D98
DD83	D99	D99
DD84	K96	K96
DD99	D99	D99
FS01	F01	F01

ICPC-3	ICPC-2	ICPC-1
FS02	F02	F02
FS03	F03	F03
FS04	F04	F04
FS05	F05	F05
FS06	F05	F05
FS07	F13	F13
FS08	F15	F15
FS09	F16	F16
FS10	F17, F18	F17, F18
FS90	F27	F27
FS99	F14, F29, F99	F14, F29, F99
FD01	F70	F70
FD02	F72	F72
FD03	F73	F73
FD04	F86	F86
FD05	F85	F85
FD25	F74	F74
FD35	F75	F75
FD36	F79	F79
FD37	F76	F76
FD55	F80	F80
FD56	F81	F81
FD65	F71	F71
FD66	F82	F82
FD67	F83	F83
FD68	F84	F84
FD69	F91	F91
FD70	F92	F92
FD71	F93	F93
FD72	F94	F94
FD73	F95	F95
FD74	F99	F99
FD99	F99	F99
GS01	Y01	Y01
GS02	Y02	Y02

ICPC-3	ICPC-2	ICPC-1
GS03	X01, Y02	X01, Y02
GS04	X18, Y16	X18, Y16
GS05	X02	X02
GS06	X03	X03
GS07	X05	X05
GS08	X06	X06
GS09	X07	X07
GS10	X08	X08
GS11	X09	X09
GS12	X10	X10
GS13	X11	X11
GS14	X12	X12
GS15	X13	X13
GS16	X14	X14
GS17	X15	X15
GS18	X16	X16
GS19	X17	X17
GS20	Y04	Y04
GS21	Y05	Y05
GS22	Y06	Y06
GS23	X04	X04
GS24	Y07	Y07
GS25	Y08	Y08
GS26	X19	X19
GS27	X20	X20
GS28	X21, Y16	X21, Y16
GS29	W15, Y10	W15, Y10
GS50	X86	X86
GS90	X21, X22	X21
GS91	X24, Y24	X24, Y24
GS92	X23, Y25	X23, Y25
GS93	X26	X26
GS94	X25, X27, Y26, Y27	X25, X27, Y26, Y27
GS99	X29, Y29	X29, Y29
GD01	X70, Y70	X70, Y70

ICPC-3	ICPC-2	ICPC-1
GD02	X71, Y71	X71, Y71
GD03	X90, Y72	X90, Y72
GD04	X73, Y99	X73, Y99
GD05	X91, Y76	X91, Y76
GD06	X74, X92	X74, X99
GD07	A78	A78
GD08	X72, Y75	X72, Y75
GD09	X74	X74
GD10	Y73	Y73
GD11	Y74	Y74
GD12	X84	X84
GD25	X75	X75
GD26	Y77	Y77
GD27	X76, Y78	X76, Y78
GD28	X77, Y78	X77
GD29	X78	X78
GD30	X79, Y79	X79, Y79
GD31	X80, Y79	X80, Y79
GD32	X81, Y79	X81, Y79
GD35	X82, Y80	X82, Y80
GD55	X83, Y84	X83, Y84
GD56	Y82	Y82
GD57	Y83	Y83
GD65	X85	X85
GD66	X87	X87
GD67	X88	X88
GD68	X89	X89
GD69	X99	X99
GD70	Y85	Y85
GD71	Y86, Y99	Y86, Y99
GD72	Y81	Y81
GD99	X99, Y99	X99, Y99
HS01	H01	H01
HS02	H02	H02
HS03	H03	H03

ICPC-3	ICPC-2	ICPC-1
HS04	H04	H04
HS05	H05	H05
HS06	H13	H13
HS90	H27	H27
HS91	H15	H15
HS99	H29	H29
HD01	H70	H70
HD02	H71	H71
HD03	H72	H72
HD04	H73	H73
HD05	H74	H74
HD25	H75	H75
HD35	H85	H85
HD36	H76	H76
HD37	H78, H79	H78, H79
HD55	H80	H80
HD65	H77	H77
HD66	H81	H81
HD67	H82	H82
HD68	H84	H84
HD69	H86	H86
HD99	H99, H83	H99, H83
KS01	K01, K02	K01, K02
KS02	K04	K04
KS03	K05	K05
KS04	K07	K07
KS50	K29	K29
KS51	K85	K85
KS52	K81	K81
KS90	K24, K25, K27	K24, K25, K27
KS99	K03, K06, K29	K03, K06, K29
KD01	K70	K70
KD02	K71	K71
KD25	K72	K72
KD35	A80	A80

ICPC-3	ICPC-2	ICPC-1
KD55	K73	K73
KD65	K74, K75	K74, K75
KD66	K76	K76
KD67	K77	K77
KD68	K78	K78
KD69	K79	K79
KD70	K80, K84	K80, K84
KD71	K83	K83
KD72	K84	K84
KD73	K86	K86
KD74	K87	K87
KD75	K88	K88
KD76	K92	K91, K92
KD77	K93	K93
KD78	K94	K94
KD79	K95	K95
KD99	K82, K99	K82, K99
LS01	L01	L01
LS02	L02	L02
LS03	L03	L03
LS04	L04	L04
LS05	L05	L05, L06
LS06	L07	L07
LS07	L08	L08
LS08	L09	L09
LS09	L10	L10
LS10	L11	L11
LS11	L12	L12
LS12	L13	L13
LS13	L14	L14
LS14	L15	L15
LS15	L16	L16
LS16	L17	L17
LS17	L18	L18
LS18	L18	L18

ICPC-3	ICPC-2	ICPC-1
LS19	L19	L19
LS20	L20	L20
LS90	L26, L27	L26, L27
LS99	L29	L29
LD01	L70	L70
LD25	L71	L71
LD26	L97	L97
LD35	L72	L72
LD36	L73	L73
LD37	L74	L74
LD38	L75	L75
LD39	L76	L76
LD45	L78, L96	L78, L96
LD46	L77	L77
LD47	L79	L79
LD48	L80	L80
LD49	L81	L81
LD55	L82	L82
LD65	L83	L83
LD66	L84	L84
LD67	L86	L86
LD68	L92	L92
LD69	L99	L99
LD70	L85	L85
LD71	L98	L98
LD72	L87	L87
LD73	L93	L93
LD74	L88	L88
LD75	T92	T92
LD76	L99	L99
LD77	L94	L94
LD78	L89	L89
LD79	L90	L90
LD80	L91	L91
LD81	L95	L95

ICPC-3	ICPC-2	ICPC-1
LD99	L99, T99	L99, T99
NS01	N01	N01
NS02	N03	N03
NS03	N04	N04
NS04	N05	N05
NS05	N05, N06	N05, N06
NS06	N07	N07
NS07	N08	N06
NS08	N16	N16
NS09	N17	N17
NS10	N18	N18
NS11	N19, P10	N19, P10
NS90	N26, N27	N26, N27
NS99	N29	N29
ND01	N70	N70
ND02	N71	N71
ND03	N72	N72
ND04	N73	N73
ND25	N74, N75, N76	N74, N75, N76
ND35	N79	N79
ND36	N80	N80
ND37	N81	N81
ND55	N85	N85
ND65	N86	N86
ND66	N87	N87
ND67	N88	N88
ND68	K89	K89
ND69	K90	K90
ND71	N89	N89
ND72	N90	N90
ND73	N95	N02
ND74	N92	N92
ND75	N91	N91
ND76	N93	N93
ND77	N94	N94

ICPC-3	ICPC-2	ICPC-1
ND99	N99, P10	N99, P10
PS01	P01	P01
PS02	P02	P02
PS03	P03	P03
PS04	P04	P04
PS05	P77	P77
PS06	P06	P06
PS07	P07, P08	P07, P08
PS08	P09	P09
PS09	P11	P11
PS10	P12	P12
PS11	P13	P13
PS12	P15	P15
PS13	P16	P16
PS14	P17	P17
PS15	P18	P18
PS16	P19	P19
PS17	P20	P20
PS18	P22	P22
PS19	P23	P23
PS20	P24	P24
PS21	Z11	Z11
PS22	P05, P25	P05, P25
PS90	P27	P27
PS99	P29	P29
PD01	P70	P70
PD02	P71	P71
PD03	P72	P72
PD04	P73	P73
PD05	P98	P98
PD06	P74, P79	P74, P79
PD07	P75, P79	P75, P79
PD08	P82	P02
PD09	P82	P02
PD10	P75	P75

ICPC-3	ICPC-2	ICPC-1
PD11	P78	P78
PD12	P76	P76
PD13	P76	P76
PD14	P77	P77
PD15	P80	P80
PD16	P81	P21
PD17	P86	T06
PD18	P85	P85
PD19	P99	P99
PD99	P99	P99
RS01	R01	R01
RS02	R02	R02
RS03	R03	R03
RS04	R04, R98	R04, R98
RS05	R04	R04
RS06	P06	P06
RS07	R05	R05
RS08	R06	R06
RS09	R07	R07
RS10	R08	R08
RS11	R09	R09
RS12	R21	R21, R22
RS13	R23	R23
RS14	R24	R24
RS15	R25	R25
RS50	R82	R82, R93
RS90	R26, R27	R26, R27
RS91	A18	--
RS99	R29	R29
RD01	R71	R71
RD02	R74	R74
RD03	R75	R75
RD04	R72, R76	R72, R76
RD05	R77	R77

ICPC-3	ICPC-2	ICPC-1
RD06	R78	R78
RD07	R80	R80
RD08	--	--
RD09	R81	R81
RD10	R83	R83
RD25	R84	R84
RD26	R85	R85
RD27	R86	R86
RD28	R92	R92
RD35	R88	R88
RD36	R87	R87
RD55	R89	R89
RD65	R97	R97
RD66	R90	R90
RD67	R79	R91
RD68	R95	R95
RD69	R96	R96
RD70	R99	R99
RD99	R99, T99	R99
SS01	S01	S01
SS02	S02	S02
SS03	S04	S04
SS04	S05	S05
SS05	S06	S06
SS06	S07	S07
SS07	S08	S08
SS08	S21	S21
SS09	S22	S22
SS10	S23	S23
SS11	S24	S24
SS90	S26, S27	S26, S27
SS99	S29	S29
SD01	S03	S03
SD02	S95	S95

ICPC-3	ICPC-2	ICPC-1
SD03	S70	S70
SD04	S71	S71
SD05	S09	S09
SD06	S10, R73	S10, R73
SD07	S11	S11
SD08	S74	S74
SD09	S74	S74
SD10	S74	S74
SD11	S75	S75
SD12	S90	S90
SD13	S72	S72
SD14	S73	S73
SD15	S84	S84
SD16	S76	S76
SD25	S77	S77
SD26	S78	S78
SD27	S82	S82
SD28	S81	S81
SD29	S79, S99	S79, S99
SD35	S16	S16
SD36	S17	S17
SD37	S18	S18
SD38	S19	S19
SD39	S12	S12
SD40	S13	S13
SD41	S14	S14
SD42	S15	S15
SD55	S83	S83
SD65	S20	S20
SD66	S80	S80
SD67	S85	S85
SD68	S86	S86
SD69	S87	S87
SD70	S88	S88
SD71	S89	S89

ICPC-3	ICPC-2	ICPC-1
SD72	S91	S91
SD73	S92	S92
SD74	S93	S93
SD75	S94	S94
SD76	S96	S96
SD77	S97	S97
SD78	S98	S98
SD80	S99	S99
SD81	S99	S99
SD82	S23	S23
SD99	S99	S99
TS01	T01	T01
TS02	T02	T02
TS03	T03	T03
TS04	T04	T04
TS05	T05	T05
TS06	T07	T07
TS07	T08	T08
TS08	T10	T10
TS09	T11	T11
TS50	T29	T29
TS51	T83	T83
TS90	T26, T27	T26, T27
TS99	T29	T29
TD01	T70	T70
TD25	T71	T71
TD26	T72	T72
TD27	T73	T73
TD55	T78	T78
TD56	T80	T80
TD65	T81	T15, T81
TD66	T82	T82
TD68	T85	T85
TD69	T86	T86
TD70	T87	T87

ICPC-3	ICPC-2	ICPC-1
TD71	T89	T90
TD72	T90	T90
TD73	T91	T91
TD74	T91	T91
TD75	T93	T93
TD99	T99	T88, T99
US01	U01	U01
US02	U02	U02
US03	U04	U04
US05	U05	U05
US06	U06	U06
US07	U07	U07
US08	U13	U13
US09	U14	U14
US10	X29, Y03	X29, Y03
US50	U71, U90, U98	U71, U90, U98
US90	U26, U27	U26, U27
US99	U29	U29
UD01	U70	U70
UD02	U71	U71
UD03	U72	U72
UD04	U71	U71
UD25	U75	U75
UD26	U76	U76
UD27	U77	U77
UD28	U78	U78
UD29	U79	U79
UD35	U80	U80
UD55	U85	U85
UD65	U88	U88
UD66	U99	U99
UD67	U95	U95
UD99	U99	U99
WS01	W01	W01
WS02	W05	W05

ICPC-3	ICPC-2	ICPC-1
WS03	W03	W03
WS04	W03	W03
WS05	W17	W17
WS06	W19	W19
WS39	W18	W18
WS50	W99	W99
WS90	W02	W02
WS91	W27	W27
WS99	W29	W20, W29
WD01	W70	W70
WD02	W71	W71
WD03	W94	W94
WD25	W72	W72
WD26	W73	W73
WD35	W75	W75
WD55	W76	W76
WD65	W82	W82
WD66	W83	W83
WD67	W78	W78
WD68	W79	W79
WD69	W80	W80
WD70	W81	W81
WD71	W84	W84
WD72	W85	W84
WD80	W90	W90
WD81	W91	W91
WD82	W92	W92
WD83	W93	W93
WD84	W95	W95, W96
WD85	W96	W77, W96
WD99	W99	W77, W99
ZC01	Z12	Z12
ZC02	Z16	Z16
ZC03	Z20	Z20
ZC04	Z10	Z10

ICPC-3	ICPC-2	ICPC-1
ZC09	Z24	Z24
ZC10	Z15	Z15
ZC11	Z19	Z19
ZC12	Z23	Z23
ZC13	Z01	Z01
ZC15	Z07	Z07
ZC16	Z05	Z05
ZC17	Z06	Z06
ZC20	Z02	Z02
ZC25	Z14	Z14
ZC26	Z18	Z18
ZC27	Z22	Z22
ZC30	Z13	Z13
ZC31	Z21	Z21
ZC35	Z25, Z13	Z25, Z13
ZC36	Z03	Z03
ZC37	Z09	Z09
ZC38	Z08	Z08
ZC39	Z10	Z10
ZC90	Z27	Z27
ZC99	Z04, Z29	Z04, Z29
-101	-30	-30
-102	-31	-31
-103	-32	-32
-104	-33	-33
-105	-34	-34
-106	-35	-35
-107	-36	-36
-108	-37	-37
-109	-38	-38
-110	-39	-39
-111	--	--
-112	-40	-40
-113	-41	-41

ICPC-3	ICPC-2	ICPC-1
-114	-42	-42
-199	-43	-43
-201	-50	-50
-202	-44	-44
-203	-45	-45
-204	-51	-51
-205	-52	-52
-206	-53	-53
-207	-54	-54
-208	-54	-54
-209	-54	-54
-210	-55	-55
-211	-56	-56
-212	-58	-58
-215	--	--
-299	-165	-165
K301	--	--
K302	--	--
P303	--	--
P304	--	--
P305	--	--
R306	--	--
R307	--	--
T308	--	--
W309	--	--
A310	--	--
A350	--	--
A351	--	--
A352	--	--
X399	--	--
-401	-60	-60
-402	-61	-61
-501	-64	-64
-502	-65	-65

ICPC-3	ICPC-2	ICPC-1
-503	-46	-46
-504	-47	-47
-505	-66	-66
-506	-67	-67
-599	-200	-200
-601	-62	-62
-602	--	--

Primary Care Functioning Scale (PCFS)

Primary Care Functioning Scale

> Radboudumc
>
> Department of Primary and Community
>
> Care
>
> Nijmegen, the Netherlands

Radboudumc

©2017, Radboudumc

Instructions

Information about this questionnaire

This questionnaire is about your functioning in daily life. Can you perform activities that are important to you? Are you not or less able to perform activities which you would like to perform better?

Instructions for filling in the questionnaire

PCFS Questionnaire:

- Choose one answer for each question. Choose the answer that best describes your situation.
- Answer the questions by putting a cross in the box corresponding with the answer of your choice.
- After each question, you are asked whether you are satisfied. Choose one answer which best describes your situation.

The questions are ***about your current situation and about how you are now.***

There are no right or wrong answers. It is about your experience and your personal situation. Good luck with filling in the questionnaire!

©2017, Radboudumc

PCFS

MULTIPLE CHOICE QUESTIONS

Date of completion of the questionnaire:
(please fill in the date)
----/----/----

©2017, Radboudumc

PHYSICAL AND MENTAL FUNCTIONS

- For each of the following questions, indicate the extent to which you experience problems (choose one answer for each question: NO problem, MILD problem, MODERATE problem, SEVERE problem, COMPLETE problem)
- For each of the following questions, indicate whether you are satisfied with this (choose one answer: Yes, Neutral or No)

1.	Feeling energetic	☐ NO problem
		☐ MILD problem
		☐ MODERATE problem
		☐ SEVERE problem
		☐ COMPLETE problem

1.a	Are you satisfied with this?	Yes ☐	Neutral ☐	No ☐

2.	Sleeping	☐ NO problem
		☐ MILD problem
		☐ MODERATE problem
		☐ SEVERE problem
		☐ COMPLETE problem

2.a	Are you satisfied with this?	Yes ☐	Neutral ☐	No ☐

3.	Feeling emotionally stable	☐ No problem
		☐ MILD problem
		☐ MODERATE problem
		☐ SEVERE problem
		☐ COMPLETE problem

3.a	Are you satisfied with this?	Yes ☐	Neutral ☐	No ☐

4.	Having generalized pain or pain in a body part	☐ No problem
		☐ MILD problem
		☐ MODERATE problem
		☐ SEVERE problem
		☐ COMPLETE problem

4.a	Are you satisfied with this?	Yes ☐	Neutral ☐	No ☐

PHYSICAL AND MENTAL FUNCTIONS

5.	Seeing		NO problem
			MILD problem
			MODERATE problem
			SEVERE problem
			COMPLETE problem

5.a	Are you satisfied with this?	Yes ☐	Neutral ☐	No ☐

6.	Hearing		NO problem
			MILD problem
			MODERATE problem
			SEVERE problem
			COMPLETE problem

6.a	Are you satisfied with this?	Yes ☐	Neutral ☐	No ☐

7.	Keeping focus and attention on a task		NO problem
			MILD problem
			MODERATE problem
			SEVERE problem
			COMPLETE problem

7.a	Are you satisfied with this?	Yes ☐	Neutral ☐	No ☐

8.	Remembering new information (memory)		NO problem
			MILD problem
			MODERATE problem
			SEVERE problem
			COMPLETE problem

8.a	Are you satisfied with this?	Yes ☐	Neutral ☐	No ☐

PHYSICAL AND MENTAL FUNCTIONS

9.	Having some exercise tolerance	NO problem
		MILD problem
		MODERATE problem
		SEVERE problem
		COMPLETE problem

9.a	Are you satisfied with this?	Yes ☐	Neutral ☐	No ☐

10.	Having a smooth joint mobility	NO problem
		MILD problem
		MODERATE problem
		SEVERE problem
		COMPLETE problem

10.a	Are you satisfied with this?	Yes ☐	Neutral ☐	No ☐

11.	Being able to use some muscle power	NO problem
		MILD problem
		MODERATE problem
		SEVERE problem
		COMPLETE problem

11.a	Are you satisfied with this?	Yes ☐	Neutral ☐	No ☐

ACTIVITIES

- For each of the following questions, indicate the extent to which you experience problems (choose one answer for each question: NO problem, MILD problem, MODERATE problem, SEVERE problem, COMPLETE problem or NOT APPLICABLE)
- For each of the following questions, indicate whether you are satisfied with this (choose one answer: Yes, Neutral or No)

12.	Solving problems	☐ NO problem
		☐ MILD problem
		☐ MODERATE problem
		☐ SEVERE problem
		☐ COMPLETE problem

| 12.a | Are you satisfied with this? | Yes ☐ | Neutral ☐ | No ☐ |

13.	Planning and carrying out daily tasks and activities	☐ NO problem
		☐ MILD problem
		☐ MODERATE problem
		☐ SEVERE problem
		☐ COMPLETE problem

| 13.a | Are you satisfied with this? | Yes ☐ | Neutral ☐ | No ☐ |

14.	Handling stress	☐ NO problem
		☐ MILD problem
		☐ MODERATE problem
		☐ SEVERE problem
		☐ COMPLETE problem

| 14.a | Are you satisfied with this? | Yes ☐ | Neutral ☐ | No ☐ |

15.	Looking after your health	☐ NO problem
		☐ MILD problem
		☐ MODERATE problem
		☐ SEVERE problem
		☐ COMPLETE problem

| 15.a | Are you satisfied with this? | Yes ☐ | Neutral ☐ | No ☐ |

ACTIVITIES

16.	Changing basic body position, e.g. standing up from a chair or bending to pick something up from the floor	☐ NO problem
		☐ MILD problem
		☐ MODERATE problem
		☐ SEVERE problem
		☐ COMPLETE problem

| 16.a | Are you satisfied with this?   Yes ☐   Neutral ☐   No ☐ |

17.	Lifting and carrying objects	☐ NO problem
		☐ MILD problem
		☐ MODERATE problem
		☐ SEVERE problem
		☐ COMPLETE problem

| 17.a | Are you satisfied with this?   Yes ☐   Neutral ☐   No ☐ |

18.	Hand and arm use, e.g. pulling, pushing, reaching and turning	☐ NO problem
		☐ MILD problem
		☐ MODERATE problem
		☐ SEVERE problem
		☐ COMPLETE problem

| 18.a | Are you satisfied with this?   Yes   ☐Neutral   No☐   ☐ |

19.	Walking a short distance	☐ NO problem
		☐ MILD problem
		☐ MODERATE problem
		☐ SEVERE problem
		☐ COMPLETE problem

| 19.a | Are you satisfied with this?   Yes ☐   Neutral ☐   No ☐ |

ACTIVITIES

20.	Climbing up and down the stairs	☐ NO problem
		☐ MILD problem
		☐ MODERATE problem
		☐ SEVERE problem
		☐ COMPLETE problem

20.a	Are you satisfied with this? Yes ☐ Neutral ☐ No ☐

21.	Driving a car or another vehicle, riding a bicycle	☐ NO problem
		☐ MILD problem
		☐ MODERATE problem
		☐ SEVERE problem
		☐ COMPLETE problem
		☐ NOT APPLICABLE

21.a	Are you satisfied with this? Yes ☐ Neutral ☐ No ☐

©2017, Radboudumc

ACTIVITIES

22.	Self-toileting	NO problem
		MILD problem
		MODERATE problem
		SEVERE problem
		COMPLETE problem

| 22.a | Are you satisfied with this? | Yes ☐ | Neutral ☐ | No ☐ |

23.	Washing yourself	NO problem
		MILD problem
		MODERATE problem
		SEVERE problem
		COMPLETE problem

| 23.a | Are you satisfied with this? | Yes ☐ | Neutral ☐ | No ☐ |

24.	Caring for your body parts without assistance (e.g. teeth, hair, fingernails and toenails)	NO problem
		MILD problem
		MODERATE problem
		SEVERE problem
		COMPLETE problem

| 24.a | Are you satisfied with this? | Yes ☐ | Neutral ☐ | No ☐ |

25.	Self-dressing	NO problem
		MILD problem
		MODERATE problem
		SEVERE problem
		COMPLETE problem

| 25.a | Are you satisfied with this? | Yes ☐ | Neutral ☐ | No ☐ |

ACTIVITIES

26.	Eating without assistance	☐ NO problem
		☐ MILD problem
		☐ MODERATE problem
		☐ SEVERE problem
		☐ COMPLETE problem

26.a	Are you satisfied with this?	Yes ☐	Neutral ☐	No ☐

27.	Doing household chores such as washing and drying clothes and garments, cleaning your living area	☐ NO problem
		☐ MILD problem
		☐ MODERATE problem
		☐ SEVERE problem
		☐ COMPLETE problem

27.a	Are you satisfied with this?	Yes ☐	Neutral ☐	No ☐

©2017, Radboudumc

PARTICIPATION

- For each of the following questions, indicate the extent to which you experience problems (choose one answer for each question: NO problem, MILD problem, MODERATE problem, SEVERE problem, COMPLETE problem or NOT APPLICABLE)
- For each of the following questions, indicate whether you are satisfied with this (choose one answer: Yes, Neutral or No)

28.	Maintaining relationships with your immediate family members	▭ NO problem
		▭ MILD problem
		▭ MODERATE problem
		▭ SEVERE problem
		▭ COMPLETE problem
		▭ I have NO immediate family

| 28.a | Are you satisfied with this? | Yes ☐ | Neutral ☐ | No ☐ |

29.	Maintaining relationships with friends, neighbours or acquaintances	▭ NO problem
		▭ MILD problem
		▭ MODERATE problem
		▭ SEVERE problem
		▭ COMPLETE problem
		▭ I have NO contact with others

| 29.a | Are you satisfied with this? | Yes ☐ | Neutral ☐ | No ☐ |

30.	Maintaining your relationship with your partner	▭ NO problem
		▭ MILD problem
		▭ MODERATE problem
		▭ SEVERE problem
		▭ COMPLETE problem
		▭ I have NO partner

| 30.a | Are you satisfied with this? | Yes ☐ | Neutral ☐ | No ☐ |

©2017, Radboudumc

PARTICIPATION

31.	Carrying out remunerative work (full-time, part-time or self-employed)	NO problem
		MILD problem
		MODERATE problem
		SEVERE problem
		COMPLETE problem
		I have NO work

| 31.a | Are you satisfied with this? | Yes ☐ | Neutral ☐ | No ☐ |

32.	Carrying out non-remunerative work (voluntary work or charity work)	NO problem
		MILD problem
		MODERATE problem
		SEVERE problem
		COMPLETE problem
		I DON'T carry out non-remunerative work

| 32.a | Are you satisfied with this? | Yes ☐ | Neutral ☐ | No ☐ |

33.	Acquiring a job, profession or work	NO problem
		MILD problem
		MODERATE problem
		SEVERE problem
		COMPLETE problem
		I am NOT looking for work

| 33.a | Are you satisfied with this? | Yes ☐ | Neutral ☐ | No ☐ |

34.	Carrying out hobbies or activities (recreation and leisure)	NO problem
		MILD problem
		MODERATE problem
		SEVERE problem
		COMPLETE problem
		I have NO hobbies

| 34.a | Are you satisfied with this? | Yes ☐ | Neutral ☐ | No ☐ |

YOUR ENVIRONMENT

- For each of the following questions, indicate the extent to which an environmental factor is a facilitator, neutral, a barrier, or not applicable to your situation. Choose one of the answers.
- For each of the following questions, indicate whether you are satisfied with this (choose one answer: Yes, Neutral or No)

35.	The medicines that I use are for me	☐ I DON'T use any medicines
		☐ A COMPLETE facilitator
		☐ A MODERATE facilitator
		☐ NO facilitator/NO barrier
		☐ A MODERATE barrier
		☐ A COMPLETE barrier

35.a	Are you satisfied with this?	Yes ☐	Neutral ☐	No ☐

36.	The aids that I use (e.g. a rollator, a walking stick, a wheelchair or a scoot mobile) are for me	☐ I DON'T use any aids
		☐ A COMPLETE facilitator
		☐ A MODERATE facilitator
		☐ NO facilitator/NO barrier
		☐ A MODERATE barrier
		☐ A COMPLETE barrier

36.a	Are you satisfied with this?	Yes ☐	Neutral ☐	No ☐

37.	The social security benefits that I have (e.g. a sickness benefit, a retirement benefit) are for me	☐ I have NO social security benefits
		☐ A COMPLETE facilitator
		☐ A MODERATE facilitator
		☐ NO facilitator/NO barrier
		☐ A MODERATE barrier
		☐ A COMPLETE barrier

37.a	Are you satisfied with this?	Yes ☐	Neutral ☐	No ☐

38.	The home care or domestic help that I receive is for me	☐ I DON'T receive any care or help
		☐ A COMPLETE facilitator
		☐ A MODERATE facilitator
		☐ NO facilitator/NO barrier
		☐ A MODERATE barrier
		☐ A COMPLETE barrier

38.a	Are you satisfied with this?	Yes ☐	Neutral ☐	No ☐

YOUR ENVIRONMENT

39.	My immediate family members are for me	I have NO immediate family members
		A COMPLETE facilitator
		A MODERATE facilitator
		NO facilitator/NO barrier
		A MODERATE barrier
		A COMPLETE barrier

39.a	Are you satisfied with this? Yes ☐ Neutral ☐ No ☐

40.	My friends are for me	I have NO friends
		A COMPLETE facilitator
		A MODERATE facilitator
		NO facilitator/NO barrier
		A MODERATE barrier
		A COMPLETE barrier

40.a	Are you satisfied with this? Yes ☐ Neutral ☐ No ☐

41.	My neighbours, acquaintances or colleagues are for me	I have NO contact with others
		A COMPLETE facilitator
		A MODERATE facilitator
		NO facilitator/NO barrier
		A MODERATE barrier
		A COMPLETE barrier

41.a	Are you satisfied with this? Yes ☐ Neutral ☐ No ☐

42.	My general practitioner (GP) is for me	I have NO contact with my GP
		A COMPLETE facilitator
		A MODERATE facilitator
		NO facilitator/NO barrier
		A MODERATE barrier
		A COMPLETE barrier

42.a	Are you satisfied with this? Yes ☐ Neutral ☐ No ☐

	YOUR ENVIRONMENT	

		☐	I have NO immediate family members
43.	The views and attitudes of my immediate family members are for me	☐	A COMPLETE facilitator
		☐	A MODERATE facilitator
		☐	NO facilitator/NO barrier
		☐	A MODERATE barrier
		☐	A COMPLETE barrier

43.a	Are you satisfied with this?	Yes ☐	Neutral ☐	No ☐

		☐	I have NO contact with my GP
44.	The views and attitudes of my general practitioner (GP) are for me	☐	A COMPLETE facilitator
		☐	A MODERATE facilitator
		☐	NO facilitator/NO barrier
		☐	A MODERATE barrier
		☐	A COMPLETE barrier

44.a	Are you satisfied with this?	Yes ☐	Neutral ☐	No ☐

©2017, Radboudumc

PERSONAL CHARACTERISTICS

- For each of the following questions, indicate to what extent you agree or disagree (choose one of the answers: COMPLETELY agree, MODERATELY AGREE, NEUTRAL, MODERATELY DISAGREE OR COMPLETELY DISAGREE)
- For each of the following questions, indicate whether you are satisfied with this (choose one answer: Yes, Neutral or No)

45.	I consider myself as an extravert person who likes to communicate with others	☐ COMPLETELY agree
		☐ MODERATELY agree
		☐ NEUTRAL
		☐ MODERATELY disagree
		☐ COMPLETELY disagree

| 45.a | Are you satisfied with this? | Yes ☐ | Neutral ☐ | No ☐ |

46.	I consider myself to be flexible, obliging and agreeable	☐ COMPLETELY agree
		☐ MODERATELY agree
		☐ NEUTRAL
		☐ MODERATELY disagree
		☐ COMPLETELY disagree

| 46.a | Are you satisfied with this? | Yes ☐ | Neutral ☐ | No ☐ |

47.	I consider myself to be conscientious, precise and careful	☐ COMPLETELY agree
		☐ MODERATELY agree
		☐ NEUTRAL
		☐ MODERATELY disagree
		☐ COMPLETELY disagree

| 47.a | Are you satisfied with this? | Yes ☐ | Neutral ☐ | No ☐ |

48.	I consider myself to be even-tempered, calm and composed	☐ COMPLETELY agree
		☐ MODERATELY agree
		☐ NEUTRAL
		☐ MODERATELY disagree
		☐ COMPLETELY disagree

| 48.a | Are you satisfied with this? | Yes ☐ | Neutral ☐ | No ☐ |

©2017, Radboudumc

PERSONAL CHARACTERISTICS

49.	I consider myself to be imaginative, interested and open to experience	☐ COMPLETELY agree
		☐ MODERATELY agree
		☐ NEUTRAL
		☐ MODERATELY disagree
		☐ COMPLETELY disagree

49.a	Are you satisfied with this?	Yes ☐	Neutral ☐	No ☐

50.	I consider myself to be cheerful, in good spirits and optimistic	☐ COMPLETELY agree
		☐ MODERATELY agree
		☐ NEUTRAL
		☐ MODERATELY disagree
		☐ COMPLETELY disagree

50.a	Are you satisfied with this?	Yes ☐	Neutral ☐	No ☐

51.	I consider myself to be confident, brave and assertive	☐ COMPLETELY agree
		☐ MODERATELY agree
		☐ NEUTRAL
		☐ MODERATELY disagree
		☐ COMPLETELY disagree

51.a	Are you satisfied with this?	Yes ☐	Neutral ☐	No ☐

52.	I consider myself to be trustworthy and honest	☐ COMPLETELY agree
		☐ MODERATELY agree
		☐ NEUTRAL
		☐ MODERATELY disagree
		☐ COMPLETELY disagree

52.a	Are you satisfied with this?	Yes ☐	Neutral ☐	No ☐

THANK YOU FOR FILLING IN OUR QUESTIONNAIRE!

©2017, Radboudumc

Alphabetical index

This index is not meant to be comprehensive, nor to be a nomenclature. It is a list only of the titles of rubrics and of inclusion terms in the rubrics. A number of inclusion show a 6-digit code for the Regional extension, and in addition show the code of the class they belong to. These comprise the synonyms and terms most commonly used in general and family practice. Users requiring a more extensive index or nomenclature can do so by using the ICPC-3 Workbench on the ICPC-3.info website. In order to maintain consistency, this should be done in cooperation with the WONCA ICPC-3 Foundation.

Abbreviations are not included in this index.

hordeolum FD02.02 — FD02
housing — 2R04
housing problem — ZC36
housing unsuited to needs ZC36.01 — ZC36
HPV-DNA test — -104
human immunodeficiency virus (HIV) screening — AP10
human papilloma virus infection — GD05
hydatidiform mole — WD26
hydradenitis SD73.02 — SD73
hydrocele GD71.00 — GD71
hydrocele or spermatocele or both — GD71
hydronephrosis — UD99
hyperactivity — PS99
hyperacusis — HS02
hyperaldosteronism — TD99
hypercholesterolaemia TD75.00 — TD75
hyperemesis — DS10
hyperemesis gravidarum WS02.00 — WS02
hyperhomocysteinemia TD99.04 — TD99
hyperinsulinism — TD70
hyperkeratosis NOS — SD99
hyperkinetic disorder — PD16
hypermenorrhoea — GS08
hypermetropia FD69.01 — FD69
hypermobility syndrome LD99.00 — LD99
hyperplasia of prostate — GD70
hypersplenism — BD99
hypertension, complicated — KD74
hypertension, uncomplicated — KD73
hypertensive heart disease — KD74
hypertensive renal disease — KD74
hypertensive retinopathy — FD67
hyperthyroidism or thyrotoxicosis — TD68
hypertriglyceridaemia TD75.01 — TD75
hypertrophic kidney — UD99
hypertrophy tonsils or adenoids or both — RD66
hyperventilation — RS04

hyphaema — FD35
hypoacusis — HS02
hypochondriasis — PD07
hypoglycaemia — TD70
hypomania — PD04
hypomenorrhoea GS07.01 — GS07
hypospadias — GD56
hypothermia — AD45
hypothyroidism or myxoedema — TD69
hysteroscopy — -112

ichthyosis — SD55
icterus — DS13
idiopathic hypertension — KD73
idiopathic hypotension — KS50
idiopathic photodermatosis SD66.02 — SD66
idiopathic thrombocytopenic — BD78
ileus DD99.01 — DD99
illiteracy ZC15.00 — ZC15
illness of child problem — ZC26
illness of parents or family member problem — ZC27
illness of partner problem — ZC25
imitating or mimicking others — 2F03
immediate family — 2R08
immediate post-traumatic stress — PS02
immune thrombocytopenic purpura BD78.02 — BD78
immunisation not carried out — AP60
immunisation or transfusion reaction — AD42
immunodeficiency disorder BD99.00 — BD99
impacted cerumen — HD66
imperforate hymen GD55.00 — GD55
impetigo — SD15
impetigo secondary to other dermatosis — SD15
implantation bleeding, a minimal haemorrhage seen at the time of implantation of the egg — WS03
impotence of organic origin — GS24
impotence or erectile dysfunction — GS24